Contents

About the author

I grew up in rural Lee County, Illinois, on a "back to basics" farm, about five miles from the nearest town. I spent much of my time checking my trap lines and hunting on the fence rows. Our nation's bi-centennial sparked an interest in history and traditional muzzleloading, and I began reading books like "The Great Adventure" by Janice Holt Giles. Soon my dream was to become a real mountain man. The next trapping season I wore a fringed leather coat, moccasins, and used my pony as a pack mule to carry traps. I roamed the fields and river banks armed with a Kentucky pistol, a Green River knife, a hand made hunting pouch with powder horn, and my hand made longrifle cradled in my arm. I was living my dream.

During the early 1980's I built a small workshop, with a forge, anvil, and old time hand tools. I taught myself to use them, and was soon making powder horns, forged knives, tomahawks, the occasional firearm, and particularly hunting pouches. As my knowledge, ambition and skills grew, other artists and collectors took notice and took me under their wing. Under their tutelage, I no longer simply re-created accouterments, I began to restore and collect original antique pouches.

I have always felt a great connection to the past. When I see and handle original pouches once used by our forefathers, I can't help but think "If only they could talk". In some ways I believe they can. Each pouch is made using techniques and materials unique to the time and place of its creation, and each pouch expresses the talents of its creator. The level of craftsmanship, type of material, decorations, amount of wear, damage, deterioration, and the accouterments that comes with a pouch tell us how it was created, used, worn, and preserved, as well as details of the person who made, or used it.

After years of examining original antique pouches, and reading their stories, I have learned to re-create the construction techniques and styles of the original makers. I also took advantage of each opportunity, and with the owner's permission I created patterns for many of the original pouches, some of which are included in this book.

To those who wish to learn the craft of traditional pouch making, I hope this book will reduce your errors, and that the techniques I learned from working with original artifacts will help you make a fine piece of historical artwork. Nothing is more satisfying than having your investment in time and materials pay off, resulting in a hunting pouch, hand made in the old way, that you are proud to use.

Dedication

I would like to thank my wife Bonnie and daughter Heather for steadily helping and encouraging me through the years as I learned this craft. I want to thank all the folks at Track of the Wolf for making this book a reality. I also want to thank everyone else that has supported my work and helped me along the way.

Sincerely…

Tim Albert
2007

Recreating the 18th Century Hunting Pouch

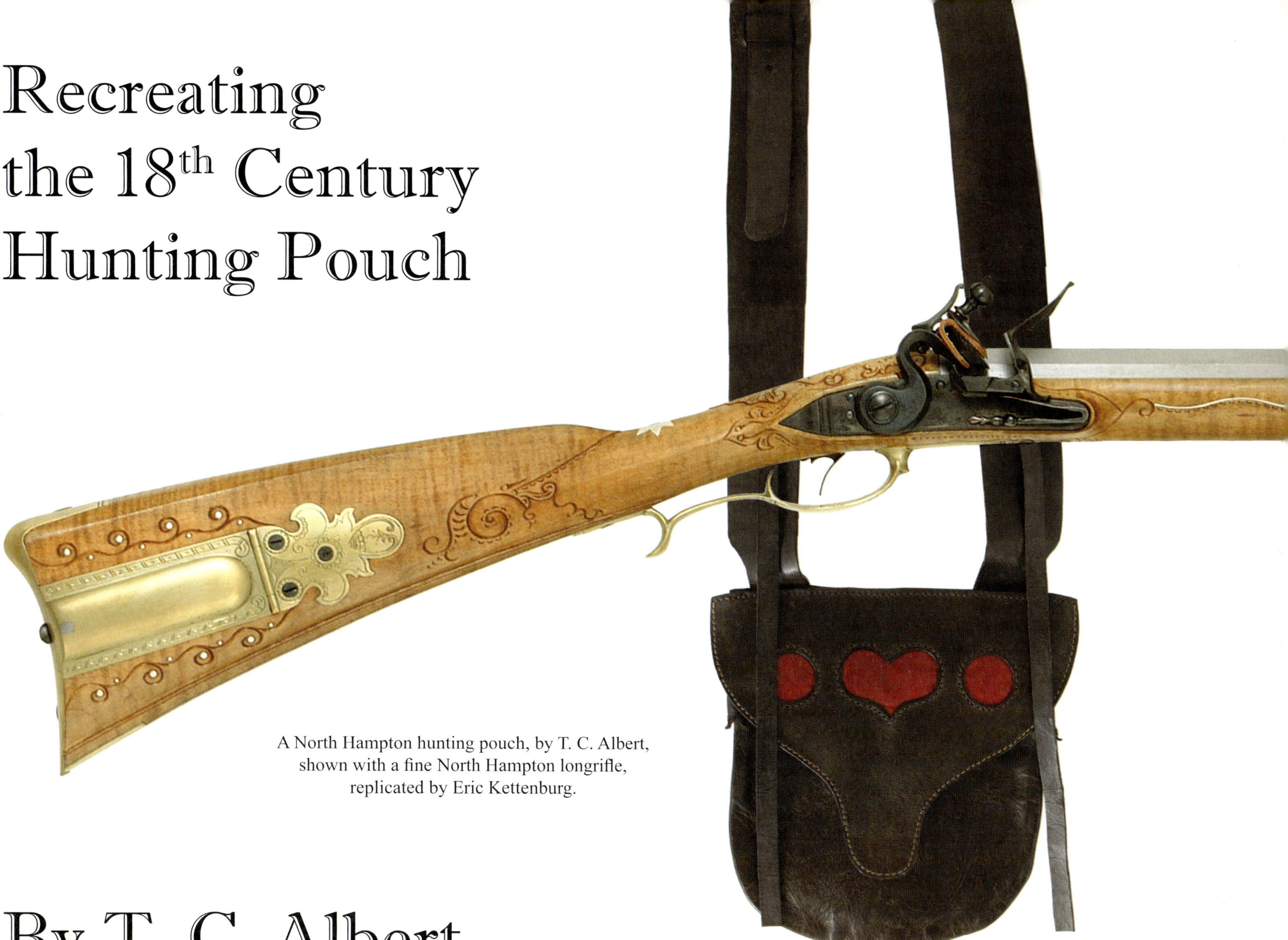

A North Hampton hunting pouch, by T. C. Albert, shown with a fine North Hampton longrifle, replicated by Eric Kettenburg.

By T. C. Albert

4-12-13

Published in the United States by:

Track of the Wolf, Inc.
18308 Joplin Street N.W.
Elk River, MN 55330-1773
www.trackofthewolf.com

Tel: 763-633-2500
Fax: 763-633-2550

ISBN 978-0-9765797-1-7

Credits:

Photography: Ethan R. Ripplinger, Ryan R. Gale

Designer: Ryan R. Gale

Editor: Ethan R. Ripplinger

Introduction

The simple Kentucky rifle of old needed few things, besides a steady hand, to do its job. Vital accessories for this gun included gun powder, lead balls of the proper size, grease and some patchin'. These staples and a few other sundries deemed necessary were traditionally carried in the leather "hunting pouch", one of the rarest of frontier artifacts.

Long rifles made of wood, brass, and iron, knives made of antler and forged steel, and powder containers fashioned from a simple cows horn all have a remarkable survival rate when compared with the leather hunting pouch. The raw skins that pioneer tanners turned into leather could only be expected to forestall the inevitable process of decay for so long. Normal wear and exposure to the elements during periods of hard use served to undermine the best preservation efforts of the finest tanneries. Given a little time or neglect, even the best pouches soon wore out quickly. Once used up, a favorite old pouch may have been consigned to the attic, barn, or wood shed out of sentiment. The conditions there wouldn't give the old pouch much of a chance of survival as the decay naturally continued. The few remaining pouches that have endured such storage are usually as dry and crumbly as a cracker and only the shriveled up shells of what they once were. Long separated from their rifles, such relics have usually been plundered for whatever horns or accoutrements of value they might have still contained and then just left to further disintegrate. Anyone lucky enough to find an original hunting pouch in any condition short of unrecognizable, would do well to study it, trying to see what secrets it has to show. Was it a fancy professionally made piece, neatly finished with exacting precision by a trained saddler? Was it a stout and sturdy plain affair made by the rough hand of the rural harness maker, maybe a little less refined than its fancy cousin, but every bit as strong or stronger? Was it a frontier pouch, stitched out in the backwoods from "home-tawed" skins, now looking and feeling for all the world like rawhide dog chew? Maybe it was homespun from an old salvaged boot, re-cobbled into a small pouch by some thrifty mountaineer using the crudest techniques. Maybe it's a mixture of all of the above, with its repairs, tears, use and abuse.

Patterning your own pouch after surviving antique examples, or using the instructions and one of the patterns contained in the following pages, is a very worthwhile and rewarding exercise. It will teach you the time-honored skills of leather working and give you the satisfaction of knowing that the pouch you have created helps to keep alive and well the art and tradition of "Recreating the 18th Century Hunting Pouch".

Tools of the trade:

Below are a few of the basic tools you will need to recreate your own hunting pouch. As your skills and goals increase, you may want to research and add more tools to your kit.

Over-stitch wheel.

Leather awl.

Buttonhole chisel.

Retractable razor knife.

Linen thread.

Heavy duty shears.

From top to bottom:

- An over-stitch wheel, used to lay out very regular and precise stitches. Found in various sizes, they will mark out different stitches per inch. I use a wheel that marks 7 stitches per inch for this project. If you buy an over-stitch wheel, I encourage you to use one, otherwise you can lay out your stitches by eye.

- A leather awl is a must for leatherwork. I prefer an awl with a diamond shaped head which cuts a small slit in the leather, instead of puncturing it like a round "ice pick" awl. A round awl has a tendency to tear the leather, if punched too close to the edge.

- A buttonhole chisel is handy for cutting slits, doing flush trim work, or cutting the odd buttonhole. Buttonhole chisels are beveled on both sides to a very sharp edge. A buttonhole chisel is wider at the tip than near the handle, unlike the parallel sides of a wood chisel.

- A razor knife with retractable disposable blades, will allow you to always have a razor sharp edge. Sharp knives can cause serious damage to you or your work. Be careful using this tool.

- Heavy duty shears for cutting leather allow your whole hand to grip, not just your thumb like normal scissors. Traditional leather purists may frown on the use of shears to cut leather. I find them perfectly suited to the task, and much quicker than a razor blade for detailed and curved pattern cuts.

- Waxed "linen" thread in white or off white is used extensively. You may want additional colors such as black and brown at your disposal. I always keep at least two or three spools on hand.

- Needles for stitching. I prefer a size '0' blunt "harness" needle. Two needles are needed at all times. After a few projects the eye will become polished, becoming very thin, cutting your thread, and eventually breaking off. Have a few extra needles on hand.

Harness needles.

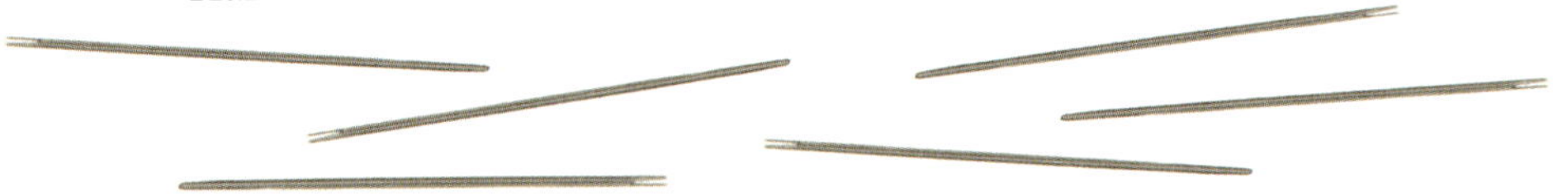

• A pair of large adjustable wing dividers, to help mark off equal distances, such as shoulder strap adjustment holes, and other reference points, is handy.

Wing dividers.

• A carpenter's adjustable scratch gauge for marking out straps. The double beam scratch gauge shown, can be set to scratch two different measurements. A single beam version will also work fine. The scratch gauge can be drawn along a straightened edge of a hide "scratching" a pre-set width into the leather. This scratch mark can be easily cut with a razor knife or shears.

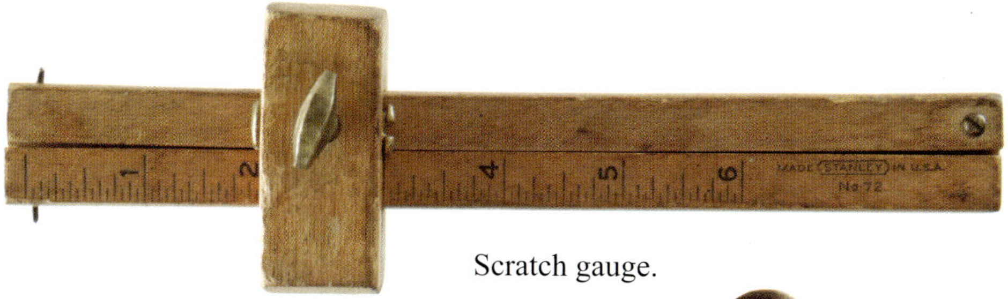

Scratch gauge.

• A slitting gauge is an old harness maker's cutting tool. The slitting gauge measures and cuts varying widths of strapping in one step. Set the desired width and pull it along the edge of the leather using the handle as a fence. The slitting blade cuts a nice parallel-sided strap. This is a difficult tool to master, especially for a left hand person. If you have a slitting gauge, use it. If not we will cover alternate techniques to create strapping.

• A rotary leather punch is used to cut rivet holes, or the holes needed for buckle tongues. Rotate to the hole size you need and squeeze the handle to punch consistently sized holes in your leather.

Most of the tools shown on these pages, or their equivalents, can be purchased new from major leather suppliers, or found used at flea markets and country auctions. You will be surprised how many of them you can find.

Slitting gauge.

Rotary leather punch.

Types of leather:

From top to bottom:
- Vegetable tanned tooling leather.
- Commercial chrome tanned vat dyed leather.
- Chrome tanned buckskin.
- Salvaged garment or "coat" leather.
- Home made, bark dyed "alum tanned" or tawed leather.

Smooth side

Rough side

On to the heart of the matter: your leather. I suggest using 3 to 4 oz. 100% vegetable tanned leather, produced commercially and available from major leather dealers. Heavier 4 to 5 oz. leather can be used, but is more difficult to work, and anything heavier should be avoided. There are other leather types available but the vegetable tanned, and homemade variety of "tawed" leathers are the best choices to accurately replicate original pouches.

Purchase as large a piece of vegetable tanned leather as possible, up to one half of a hide (approximately 24 sq. feet). A whole side will let you make all of your straps to length without splices, and will make 4 or 5 complete pouches. Many leather dealers will refer to vegetable tanned leather as "tooling" leather. The leather for your pouch should have a smooth side and a rough side. You do not want to use suede leather, or what is called an "economy split", which has two "rough" sides. The smooth side of the leather was the animal's hair side. The smooth side will eventually become the outside of your pouch, although it is sewn "rough" side out, it will be turned smooth side out as we progress.

Leather will be your biggest investment. In a pinch, any of the chrome or salvaged leathers can be used. However, those leathers will not dye or antique as well as vegetable tanned leather when using the processes described.

Again, let me stress that using the proper tools and leather will impact your finished pouch. If you can not get the recommended item, use what you have, what you can make or scrounge. The skills you acquire by "making do" will serve you well, and the hunting pouch you produce will be all the more uniquely yours.

Leather thickness chart.

- 2 - 3 oz.
- 3 - 4 oz.
- 4 - 5 oz.
- 5 - 6 oz.
- 6 - 7 oz.
- 7 - 8 oz.
- 8 - 9 oz.
- 9 - 10 oz.
- 10 - 11 oz.

Setting up shop:

Leather work lends itself to small spaces, like the corner of the family room, you do not need a large workshop. All you need is a place to sit, a place to keep your tools organized, and now and then a bit of open floor space to roll out and cut pieces of leather. Occasionally you will need a warm water source, and a small bucket or container.

Here you can see my typical work space. I always have a good bright window or electric light at my back and overhead. Fighting shadows for hours at a time is very hard on the eyes, and the quality of your work will suffer.

Notice the board across my lap. I use this as a cutting board, straight edge, and soft surface to poke my awl holes against. I prefer a simple board rather than the traditional stitching pony or "clam". Use a smooth and fairly knot free plank of soft pine or cedar. Hardwood cutting boards or butcher blocks can snap your awl tips in an instant and should be avoided.

Your work space can be very compact, which lends itself easily to today's modern living. In times past, a traveling cobbler carried his shop on his back as he visited family after family. Only a few tools, good light, a little space, and leather were all that was needed to "turn a fair bit of cordwaining".

When applying dyes and antiquing agents I would strongly recommend the use of proper ventilation, or better yet, do these steps outdoors. The stains and dyes can be messy, and the fumes can be harmful.

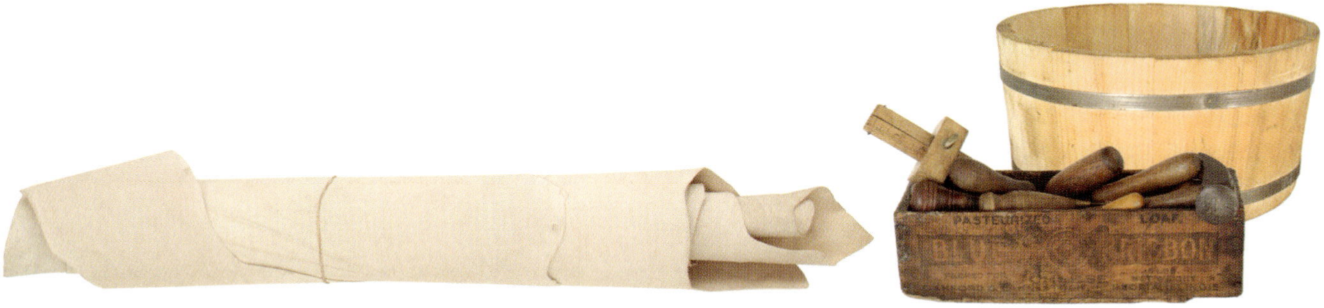

Essential skills to practice before you begin:

1. Skive the overlapping edges with a sharp knife.

2. Scrape the tapered edge flat with the edge of a knife.

3. Two pieces skived on opposite sides ready to splice.

Before starting your pouch, I strongly recommend practicing some basic techniques. A good place to start is with a traditional harness splice. By beveling or "skiving" two pieces of strapping (one on the smooth side and one on the rough side), and splicing them together a strong joint of uniform thickness is formed.

Figure 1. Begin with two short straps of leather approximately 3/4" wide. Using a sharp razor knife, carefully skive off a piece of leather from the tip of the one of the straps. Start about 3/4" from the end on the smooth side. Start your cut, going all the way across the strap, cutting deeper and deeper until you are touching your cutting board just as you get to the end of the leather. Your strap should now have a very uniformly taper cut on the end.

Figure 2. If you have an uneven cut, you can gently scrape it smooth with the edge of a sharp knife. This will level off the skive if there is any unevenness.

Figure 3. Repeat the process on the rough side of the second strap. The idea is that two tapered ends will overlay each other. If tapered neatly, the splice will be the same thickness as a single piece.

4. Overlap the two skived ends.

5. Dampen the skived ends in water.

Figure 4. Overlap the two skived straps with both smooth sides up. Always allow at least 3/4" of overlap on a splice joint.

Figure 5. Once you are satisfied with the fit of the skived ends, dip them into warm water getting 1" or more wet on both straps. The ends only need to be damp, and not soaked. It is easier to work and stitch vegetable tanned leather when slightly damp.

Figure 6. Overlap the pieces, make sure the splice is lined up and both smooth sides are face up. Lay the strap on your board, and carefully lay out several stitches on each side by gently rolling your over-stitch wheel across the leather about 1/8" from the edge. Notice how I use my finger to maintain a consistent offset from the edge. If you don't have a wheel, lay-out your holes by eye. Make at least one stitch beyond the ends of your overlap. Do not make your stitches too close to the edge, else your awl or needles may split the edge of the leather.

I do not use glue. Glue blotches on the leather will stand out when you dye it. Pulling the needle and thread through a freshly glued joint guarantees that glue will get on those surfaces, no matter how carefully it is applied.

6. Lay out the stitches with an over-stitch wheel.

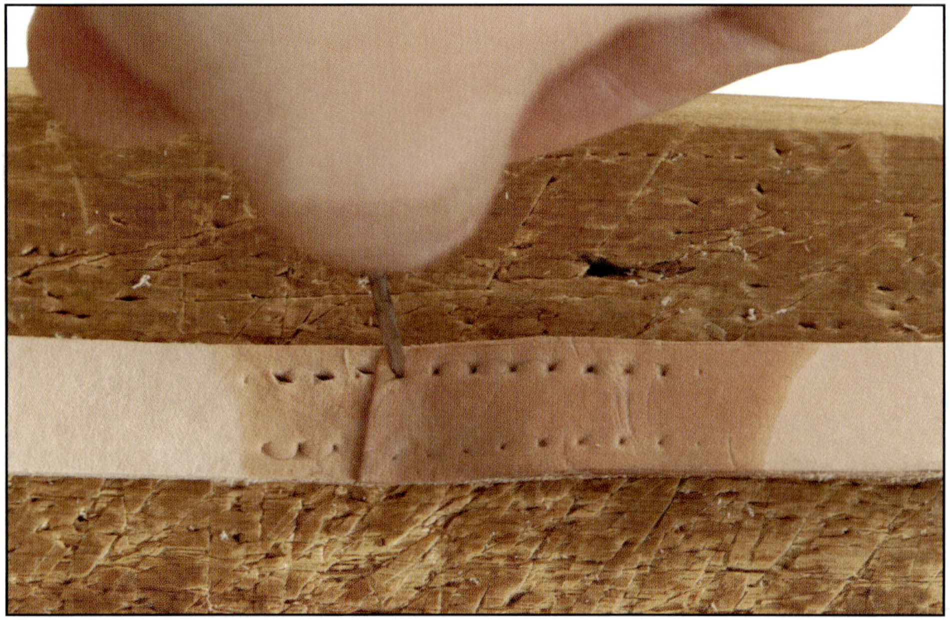

7. Punch holes with an awl.

8. Threading a needle.

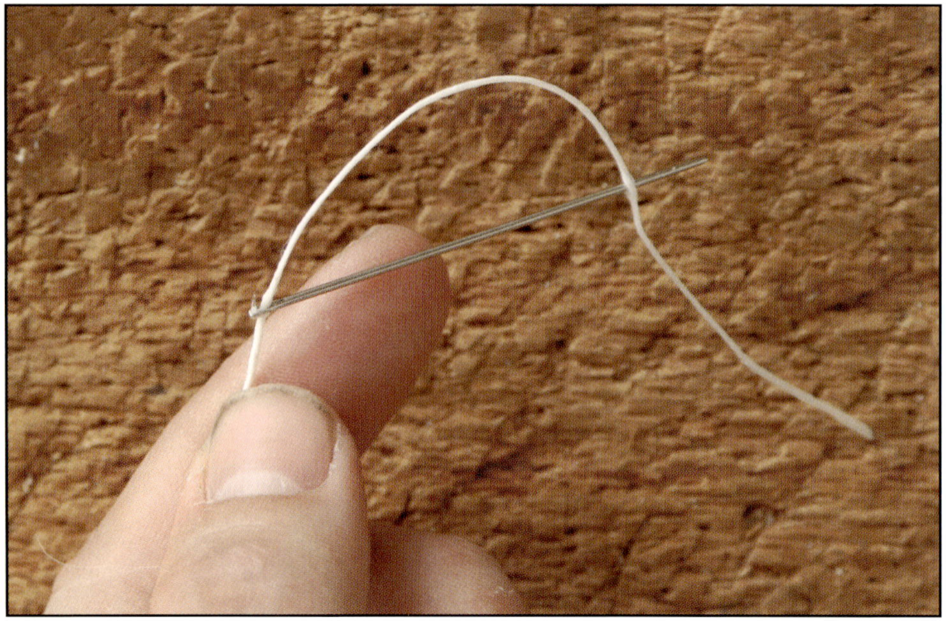

9. Split thread with needle to secure.

Figure 7. Using an awl, punch a hole in every mark laid out by the over-stitch wheel. Punch completely through both pieces of leather. When using a diamond shaped awl, keep the awl aligned so the holes are oriented in the same direction. This will keep your stitches looking uniform. Do not allow your pieces to slip around or move, else your holes will not line up properly.

Figure 8. Now its time to throw a lock hitch on your needles. Take a length of waxed linen cord and flatten the end by pulling it between your thumb and finger nail. Thread it through the eye of a harness needle. Pull a length of thread about twice as long as your needle through the eye. I call this the “tail”.

Figure 9. Halfway up the tail poke the needle through the center of the thread, splitting the thread and pushing the needle completely through the thread. Hold the tip of the needle in one hand and gently pull the long end of the thread to lock it. You now have a securely threaded needle with a roughly two inch long tail. You do not need to lock hitch your thread and needle, but it helps keep the thread from being accidentally pulled back through the eye.

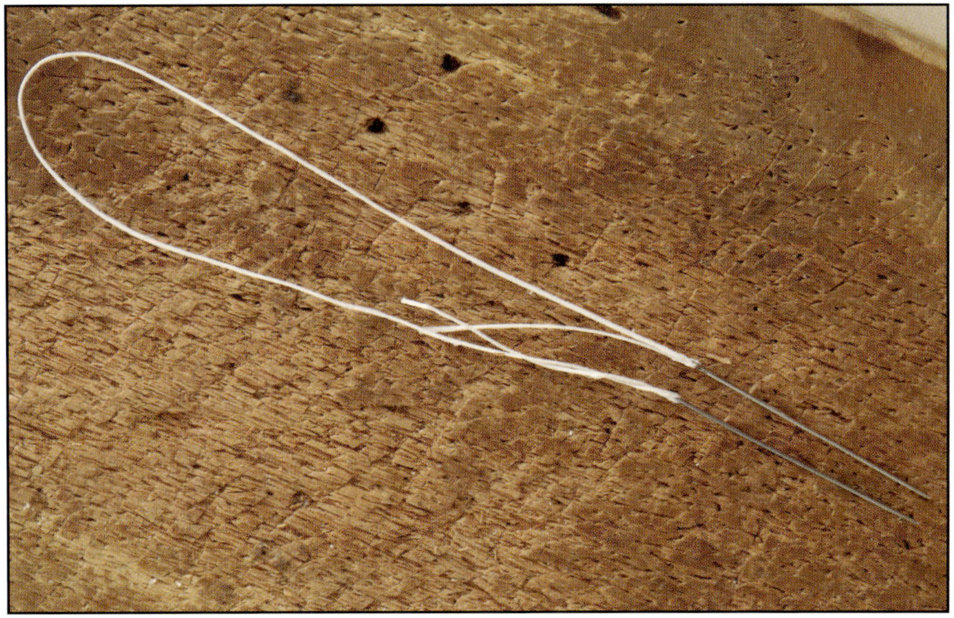

10. One thread with two needles.

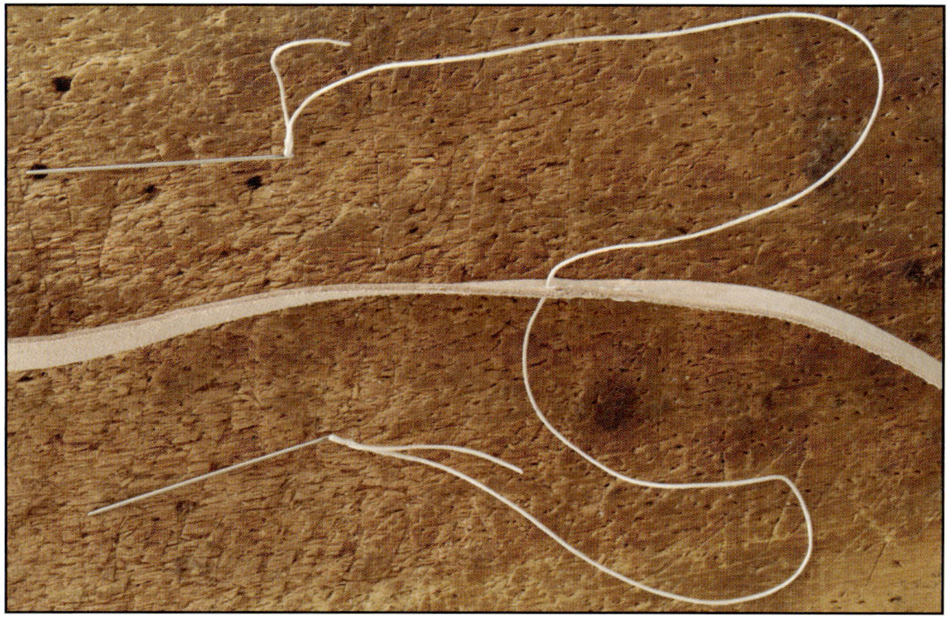

11. Pass through the first hole and center the thread.

Figure 10. Repeat the threading process on the other end of your thread with an additional needle. Your thread should have a needle on each end. One thread will be used to stitch through both sides of the leather. Make your thread length at least double the length you need to stitch. For a dozen or so stitches, use 8" or 10" of thread. On longer runs around the body of the pouch, a 36" or longer thread is preferred.

Figure 11. Make sure the leather is still a little damp. Push a needle through one of the corner awl holes. Pull half of the thread through the hole, centering it.

Figure 12. Push the needle on the smooth side through the next hole in the run and gently pull it taught, but not so tight that you off center the thread. Next push the second needle up through the same hole, but from the rough side, being careful not to catch any strands of thread from the previous needle. Pull both threads taught using both hands, pulling in opposite directions.

12. Pass both needles through one hole and pull taught.

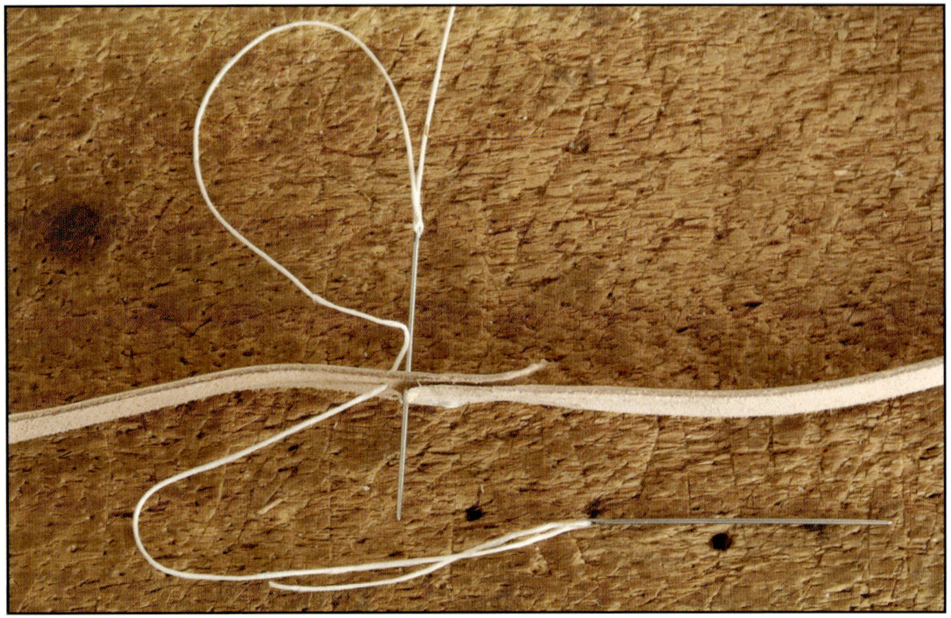

13. Continue stitching.

14. Stitch over the splice.

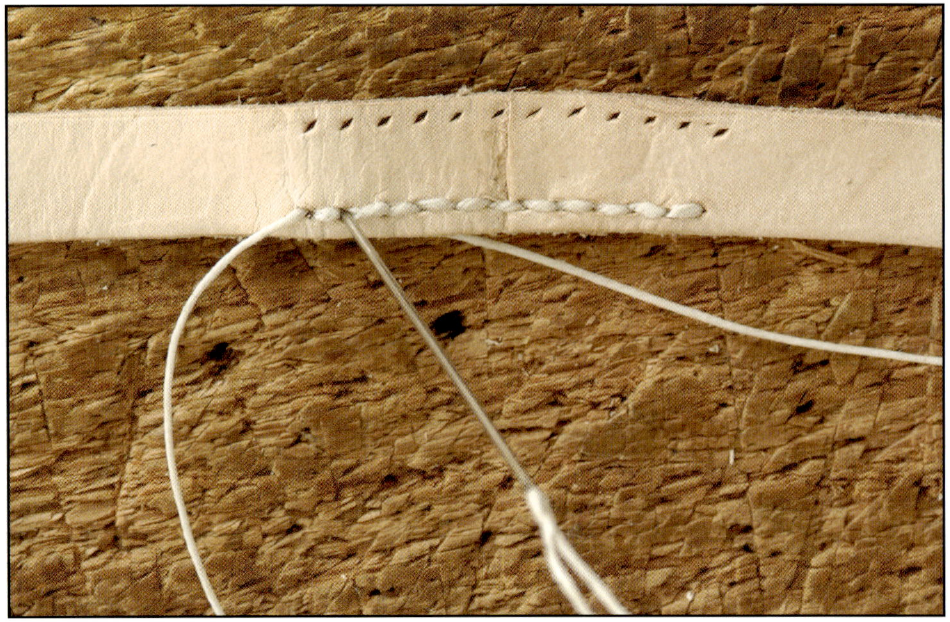

15. Back stitch twice to secure.

Figure 13. Continue stitching the leather along the run of holes. The leather should remain damp. This allows your stitches to pull a bit below the leather's surface on the smooth side, and stay tight even after the leather dries.

Damp vegetable tanned leather can be decorated using metal tooling stamps. The stamp marks will be permanent when dry, a trait called memory. Scratching damp leather with your finger nail, or resting a tool on it will also leave a permanent mark. Laying iron tools on damp leather can also leave a permanent black stain as the tannin in the leather reacts with the iron. This black stain may not compliment your finished piece.

Figure 14. When you come to the splice, continue stitching over it as shown.

Figure 15. At the end of the stitching run, back stitch at least two full stitches with both needles.

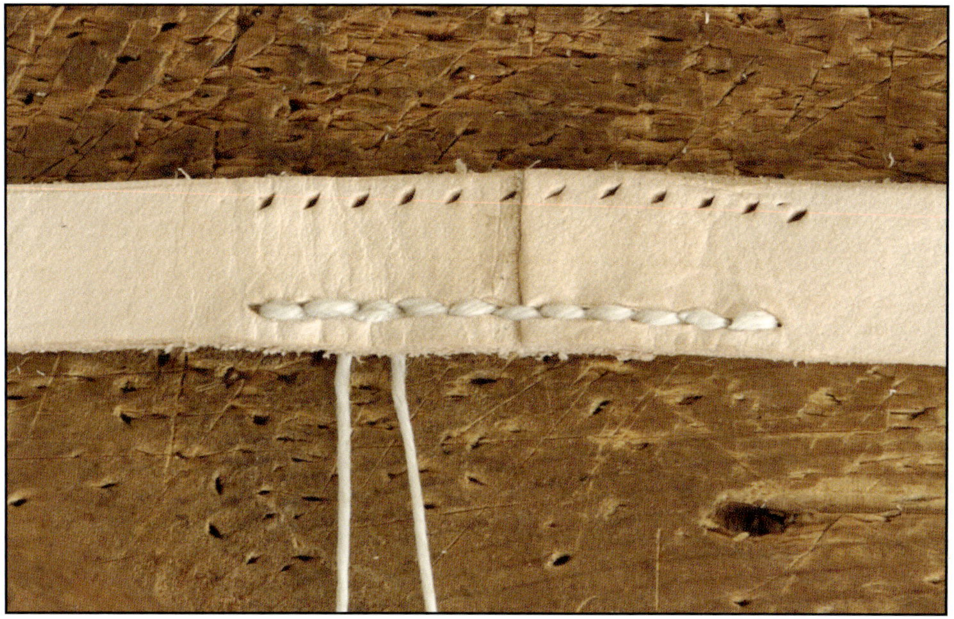

16. A finished back stitch.

17. Cut the thread flush with the leather, no need to tie it.

Figure 16. Push the top needle back through one last time so both needles are now on the rough side.

Figure 17. Trim the threads flush with your knife on the rough side of the strap. No loose ends will show on the smooth or "finished" side. There is no need to tie a knot.

Figure 18. Stitch the opposite side repeating the same steps. Start by lock hitching two needles on a fresh length of thread. Finish the run with two back stitches, and flush trim the thread on the rough side. Running your over-stitch wheel back and forth over the top of your finished stitches will make them look more crisp. That's why it is called an over-stitch wheel.

Looking down on the smooth side, your completed harness splice should look like the finished one shown. This is a very strong splice, even on thin leather or narrow straps. When done neatly it will add a very nice touch to your work. If splicing two pieces to be used for a shoulder strap, try to locate the splice on the wearer's back where it will be less visible. This may not always be possible, but it makes for a better looking finished pouch.

18. Continue the process on the other side.

The original hunting pouch:

Just as the guns of Lehigh Valley, Bethlehem, and Allentown, Pennsylvania represent some of the finest “Golden Age” long-rifles, this North Hampton County hunting pouch represents the zenith of traditional pouch making during the same era. This pouch combines the free flowing simplicity of traditional Pennsylvania folk art with the unyielding precision of the Swiss and German immigrants’ craftsmanship.

Made between 1810 and 1819, the pouch measures 8-3/4” wide by 10” long with a 6-1/2” long flap. The flap has two 1-1/4” diameter cloth inlays, and a 2-3/8” wide heart. The fine stitches on the flap are spaced 1/8” apart.

The powder horn measures 13-1/4” long, and originally had a separate screw tip, now missing. The plug is 2-3/8” diameter and is retained by wooden pegs. The horn has two iron staples for the leather straps.

Original North Hampton hunting pouch.
Jack Vye collection.

The original pouch accompanied a rifle made by Nathan Moll, gunsmith and working member of Pennsylvania's famous "gun making Moll" family. Nathan was also a close neighbor to such other famous gun makers as the Kuntz, Newhart and Rupp families. This fine pouch would be at home with any of the classic longrifles by these North Hampton County gunsmiths. It serves as a fine example of the quality and style of the hunting pouch created to accompany the longrifle. The hunting pouch is an early American folk art form, with an unique style that compliments the longrifle it accompanies.

The shoulder strap is made of 2" wide leather, folded in half, sewn along the raw edge. The strap has $^{5}/_{8}$" wide leather adjustment straps, with a thin iron buckle.

The corner seam reinforcements are 3" long. The bag is sewn with welted seams, and has a 1-$^{3}/_{8}$" wide tapered gusset sewn into the bottom.

Cut out the pieces:

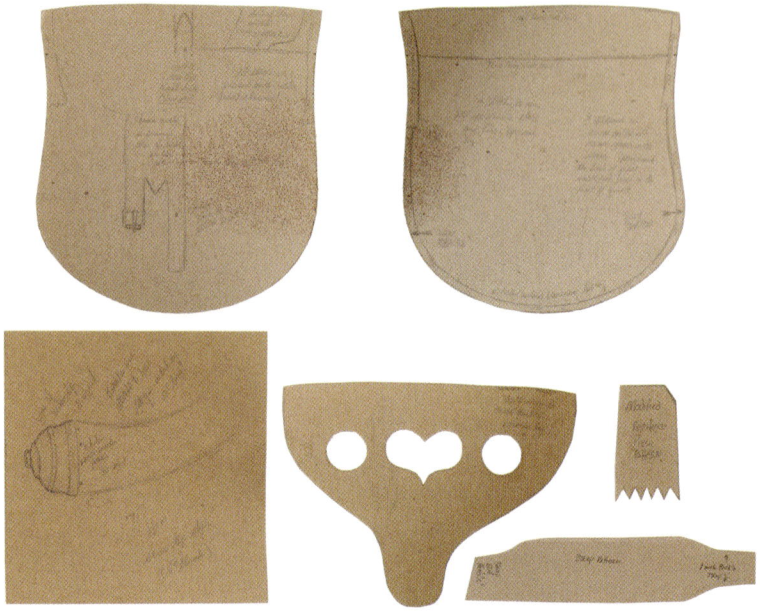

19. The pattern pieces.

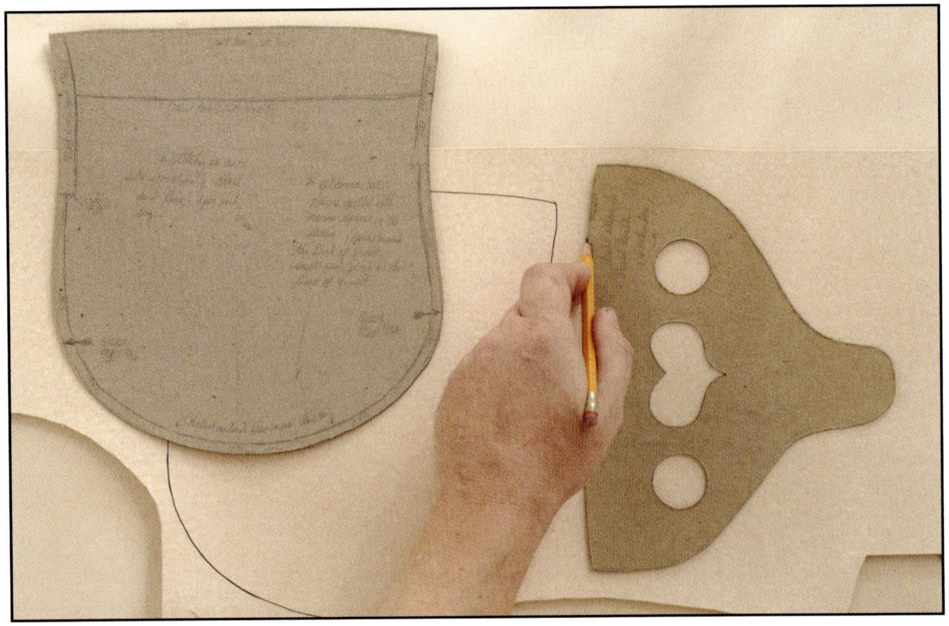

20. Trace the pattern onto the leather with a pencil.

A pattern or template of your pouch will provide two benefits. First, you can arrange the pattern pieces on your leather to minimize the amount of waste. Second, you can easily make the same bag again, or improve on your first attempt.

To make a pattern, study the actual pouch you want to copy. Take as many measurements, direct tracings, and photos as you can. Reference books of original pouches are a good source for ideas. These photos often lack any real size dimensions, so you will need to estimate size. An average antique pouch is not much larger than a sheet of 8-1/2” by 11” paper. Since reference book photos will not show everything, you will need an educated guess on the parts that do not show. Does the profile of the body behind the flap follow the general line of the pouch, or is it slightly different? None of the original pouches I’ve replicated or repaired were perfectly symmetrical side to side. If your pattern is asymmetrical, trace one side then reverse it to trace the other side. Draw your pattern onto paper first, until you are satisfied with the pouch layout.

Figure 19. Transfer your paper pattern drawings to heavy cardboard, cut them out, and you will have a set of durable pattern pieces. The cardboard pattern is a convenient place to jot down any construction notes or details you want to remember. The pattern shown was copied from the original North Hampton County hunting pouch. The tapered gusset panel bottom has been left out for simplicity on this example. It retains the dimensions and profile of the original pouch. The advanced techniques for making and installing the tapered gusset panel are discussed in a later chapter.

The pattern for this pouch, as well as several others,
are found in the last chapter of this book.
Photocopy or trace them on a
separate piece of paper
before cutting out.

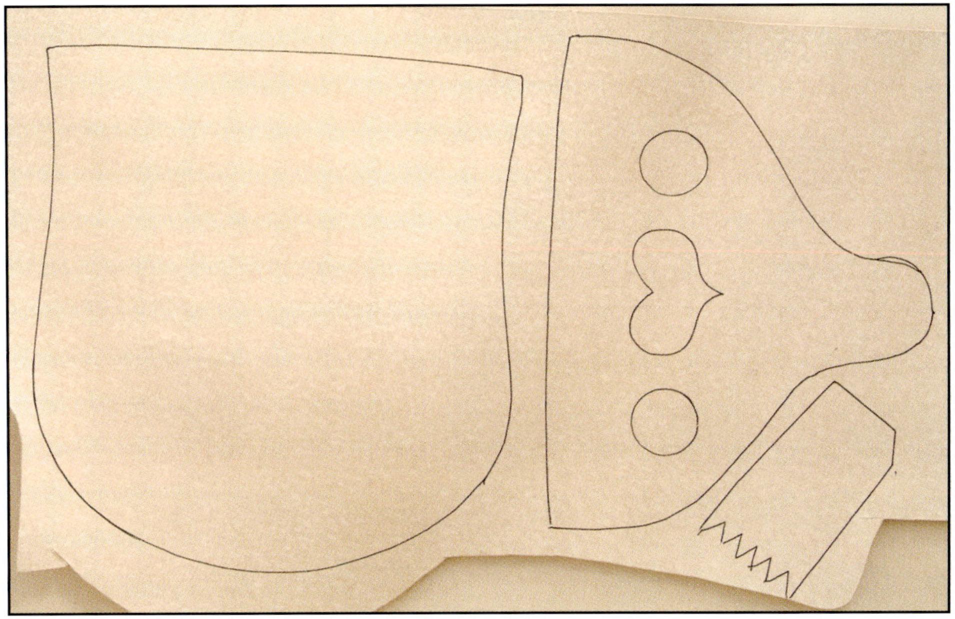

21. The traced pieces.

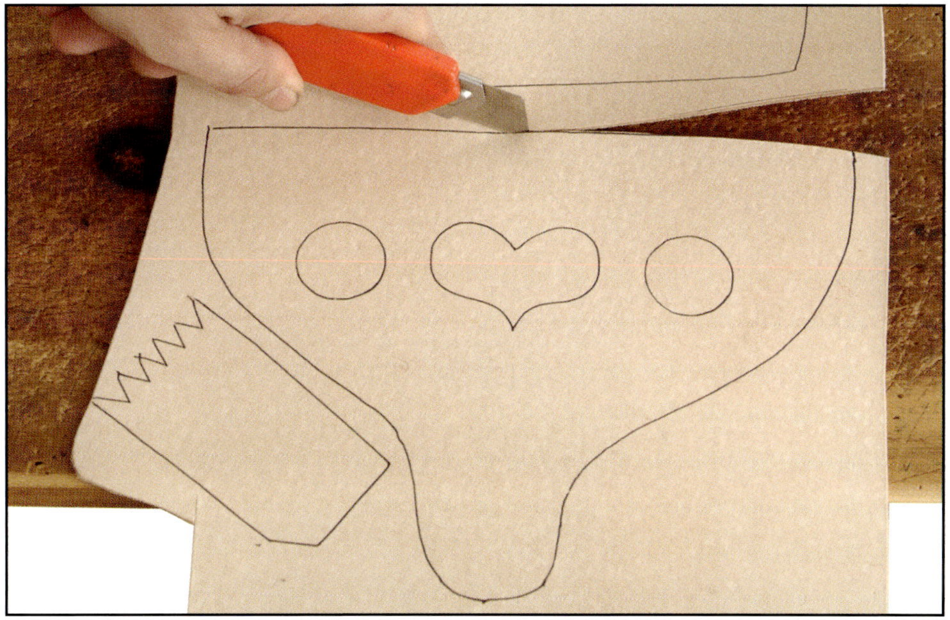

22. Cut out the pieces with a sharp knife or shears.

Figure 20-21. Using a soft pencil, trace your pattern pieces on the smooth side of the leather. Work from the uneven or belly edge of the hide for the pouch and flap pieces. Save the nice straight backbone edge for your strapping. Never use a pen to trace your patterns. Pen marks on the surface of your pouch will look unappealing. Maximize your leather by laying out pieces close to each other. Remember there is a front and back to the pouch, so be sure to trace out a front and a back, not two fronts or two backs.

Figure 22. Cut out your pieces. Use shears, or a sharp razor knife. There is really no great advantage to either. The razor knife can be faster, but can slip and ruin the piece if your not careful. Place a cutting board under the leather when cutting with a razor knife.

Figure 23. The focal point of the North Hampton County pouch is the decorative inlays. Placing the flap on a cutting board, cut out the heart and circles for the inlays. Carefully cut each with a sharp knife. Make these cuts as perfect as possible. It is hard to hide an incorrect cut on the front of the pouch.

23. Cut out the decorative inlays with a knife.

24. Scribe a line for the shoulder strap with a scratch gauge, or ruler and pencil.

25. Cut out the shoulder strap with a knife or shears.

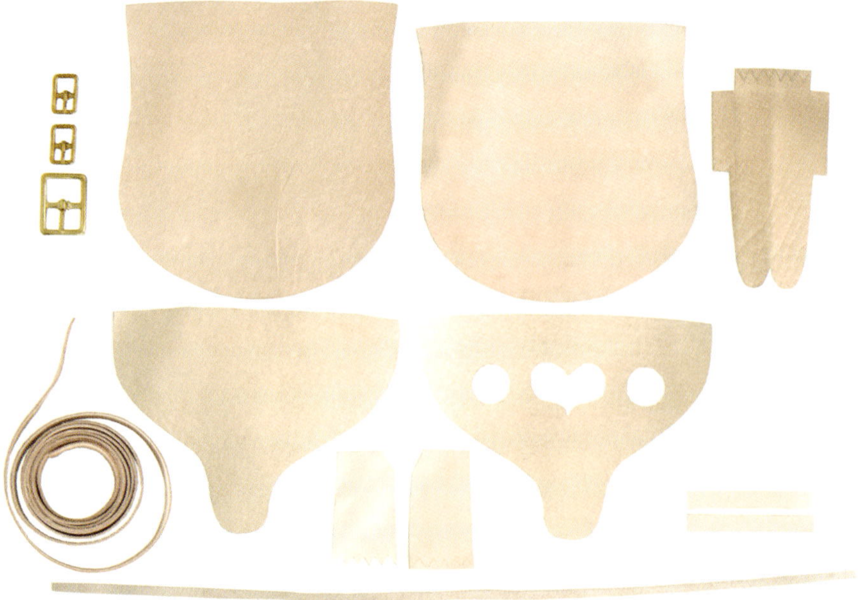

26. The finished pieces, straps, and buckles.

Figure 24. Using a long straight edge and a pencil, draw a straight line to mark an edge of the strap. The entire length of the hide may be required to make a strap without splicing, a typical strap is 73" or longer. This pouch will use 20" for the buckle strap, 53" for the adjustment strap. Cut a straight edge with a knife or shears. Set your scratch gauge to the desired width, 2" on average. Resting the straight edge of the gauge against the straight edge of the leather, scribe a line down the entire length of the hide. You will find it easier to have a board under the edge of the hide to guide the gauge. The leather may not be stiff enough unsupported, and your mark may wander if the hide curls. If you do not have a scribe, measure and mark the width of the strap with a ruler or straight edge and a pencil. Use the straight edge of the leather as a reference point.

Figure 25. Cut the 2" strap with shears following the scribed line carefully.

Figure 26. The complete set of pieces: from top left, notice the three brass center bar buckles (one 1" and two 1/2"), back of the body, front of the body, and the patch knife sheath. Bottom left: shoulder strap, back of the flap, two reinforcements, front of the flap with piercing, and the short powder horn buckle keepers. At the bottom, a 3/8" wide by 24" long strip for the body welt.

Constructing the flap:

27. Skive the top edge on the rough side of the pierced flap.

28. Dampen the edges of both flaps with water.

Figure 27. Slightly skive the top edge on the rough side of the pierced flap. Be careful not to cut all the way through to the smooth side, or make an uneven edge. Skiving the top edge will help create less bulk when you sew the flap to the body of the pouch.

Figure 28-29. Dampen the edges of both flap pieces with warm water. Make a mark 3/8" down from the top edge of the smooth side of the pierced flap piece, with a wing divider or a ruler. Repeat on the opposite side. This is where your stitches will begin and end.

Forged awls like this one were popular trade items during the 18th and 19th Century. The double points allow one end to be inserted into a wood or bone handle, or to be used for punching holes.

29. Measure the distance from the flap edge for stitching.

30. Lay out the stitches using an over-stitch wheel between the two marks.

31. The stitching guides shown around the edge.

32. Line up both flap pieces and punch a hole in each corner.

Figure 30. Lay out the stitch marks along the entire outer edge of the flap piece with an over-stitch wheel. Start and end at the marks you laid out with the wing divider. Do not mark the top edge. This will be stitched together later. Stay uniformly close to the edge as you go. I typically make the center of my stitches 1/8" from the edge. You may find it easier to control the distance by using your extended thumb as a fence by running it along the edge of the flap.

Figure 31. Place the front flap with the edge marked by the over-stitch wheel over the back of the flap, rough side to rough side.

Figure 32. Align the front and back of the flap. Punch a hole with an awl through both pieces at the first stitch mark.

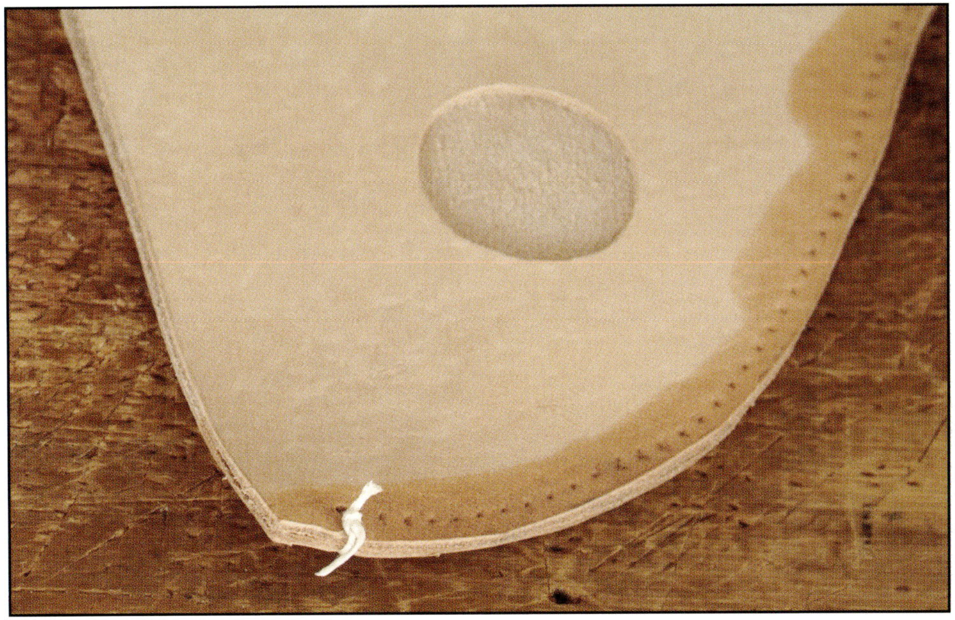

33. Tie off the corners to secure while punching and stitching.

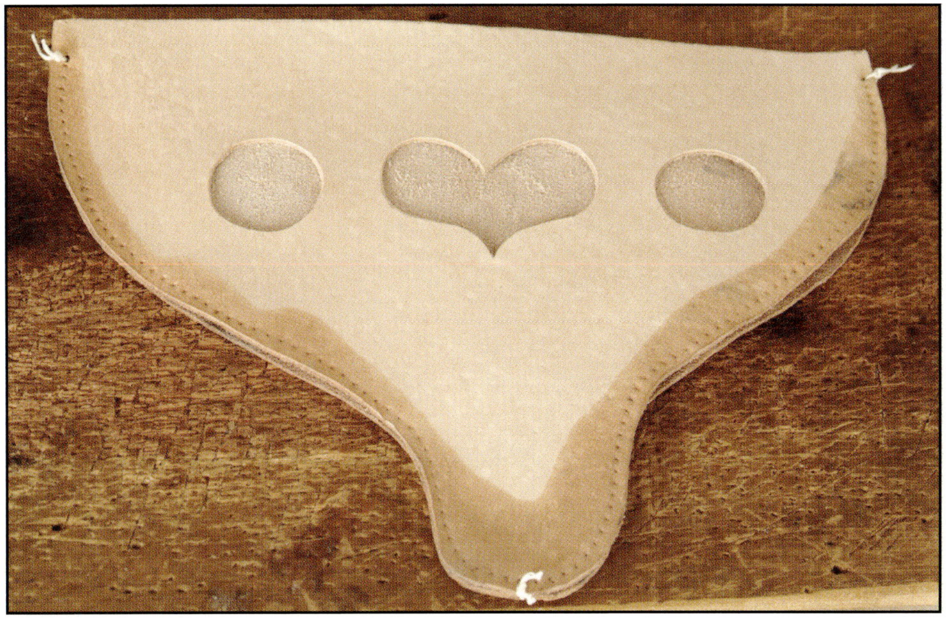

34. The flap with stitching guides and corners tied.

Figure 33-34. Tie on some keeper stitches. Thread a short piece through the awl hole, tie an overhand knot, and trim the tails closely. This keeps the flaps from moving while being stitched together. No matter how careful, without keeper stitches, there is a tendency for things to come out uneven. Add a second keeper stitch in the center of the flap, and a third at the opposite corner.

Figure 35. Punch the holes laid out by the over-stitch wheel. Go completely through both pieces of leather. Keep the slits from a diamond point awl angled in the same direction, as you go around the edge of the flap. This will keep your stitches uniform.

Figure 36. Thread a pair of needles with a long length of thread. I am using black thread for contrast, but you can use any color. Center the thread on the first hole with the keeper stitch intact, and begin saddle stitching the flap together.

Figure 37. Saddle stitch around the edge of the flap, ending with at least two back stitches, as you did on the practice splices.

35. Punch the stitching holes with an awl.

36. Thread two needles and begin saddle stitching the flap.

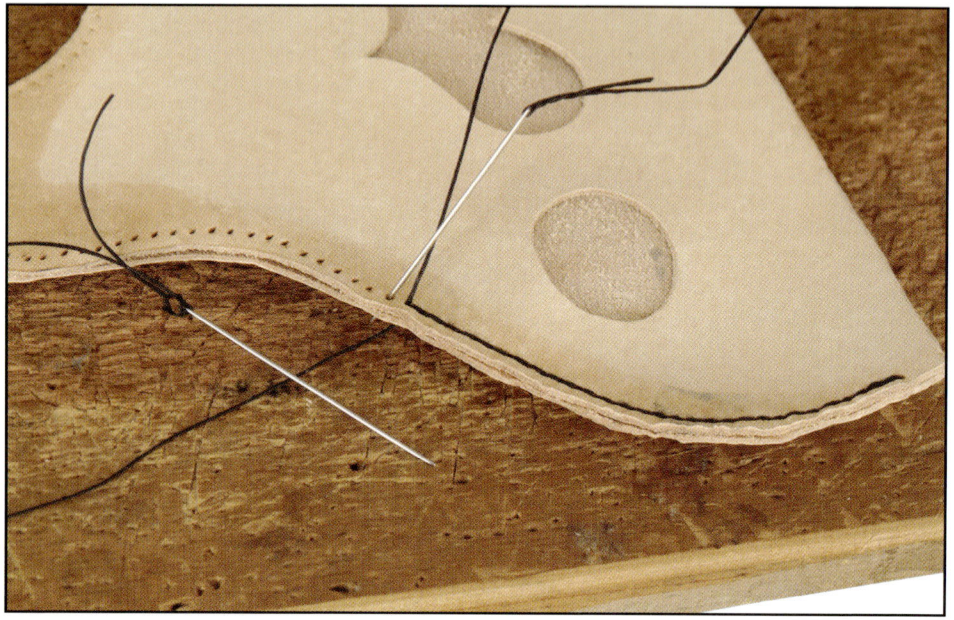

37. Saddle stitch the flap pieces together.

38. Prepare the leather dye.

Figure 38. On a pierced flap, it is best to dye the inside edges of the piercing before inserting the fabric and stitching around the inlays. It will be impossible to dye the leather without ruining the inlay fabric once the inlay is stitched. I use a home made harness dye, consisting of a quart of pure cider vinegar plus a big handful of old rusty iron. The dye is then allowed to brew for a week or more in a sealed container. A Mason jar works well. Only pure rusty iron should be used: old nails, fence wire, or similar sources. Modern alloyed steel will not yield the same results. As it brews your dye may be cloudy one day, then clear the next. It may be foamy on the top of the jar, or be full of sediment and sludge at the bottom. The only way to tell if the dye is ready, is to actually swab some on a test patch of scrap leather. If the dye is ready, it will immediately turn the leather an intense blue-black, then turning to a charcoal gray as it dries. The brewing may take some time depending on the vinegar, the temperature, and the iron. Prepare this mixture ahead of time if you want to use this type of dye on your pouch. Another alternative is to use a commercial leather dye.

39. Dye the edges around each piercing.

Figure 39. Using a small brush, carefully apply dye around the raw edges of the inlays. The natural dye will slowly darken as it dries.

Figure 40-41. Allow the flap to dry completely. Cut your inlay material considerably larger than the piercing so there will be enough to stitch in place. Place the inlay facing fabric between the two flap pieces. I'm using a piece of red velvet, as used on the original pouch. You can use the fabric of your choice: wool, linen, and leather of different colors and textures make good inlays.

Add nails and old iron parts to cider vinegar, to create leather dye.

40. The flap stitched together with edges of each piercing dyed.

41. Insert the inlay facing material.

42. Dampen the edges of the cut-outs with warm water.

43. Lay out the stitches with an over-stitch wheel.

44. Punch the holes with an awl.

Figure 42. Using warm water and your finger, dampen the edges of the inlay piercing on the front of the flap.

Figure 43. Lay out the stitches around the three piercings with an over-stitch wheel. Use your finger as a guide to keep a consistent distance from the edge of the inlay. I try to stay 1/8" from the edge.

Figure 44. Punch the awl holes through both layers of leather. Pay attention to the angle of the awl holes when using a diamond awl. For the heart inlay, punch a hole at the tip and the top where the lobes meet. This keeps the stitches centered, the over-stitch wheel may not meet at these points.

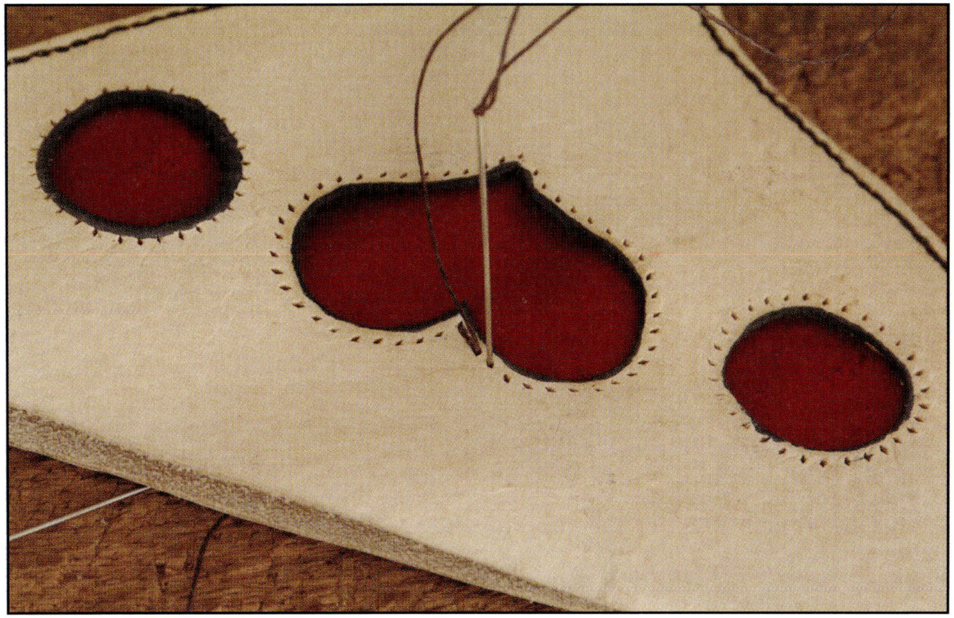

45. Saddle stitch the flaps and facing material together.

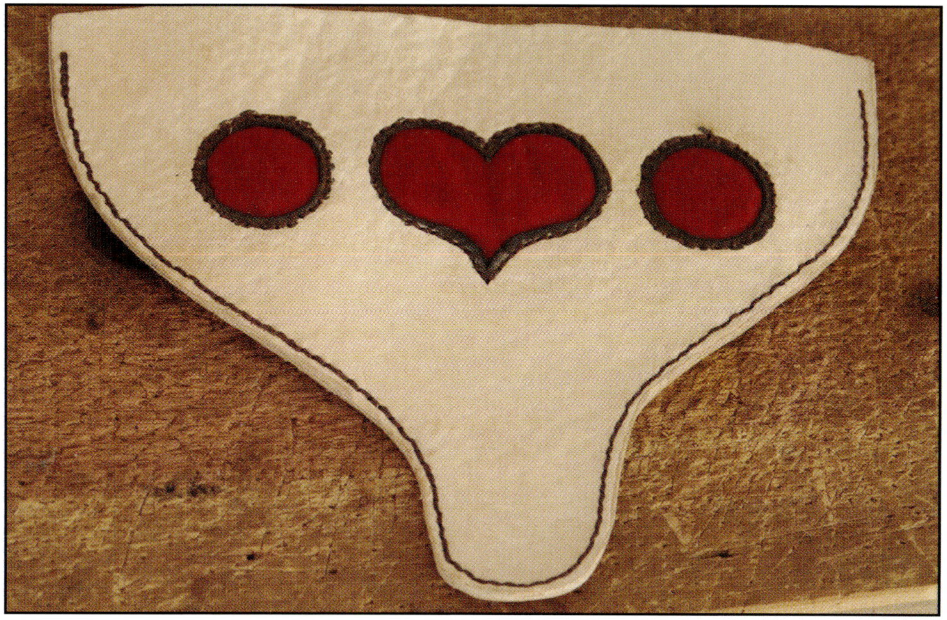

46. The flap with facing material stitched in place.

Figure 45-46. Saddle stitch around the heart with thread and double needles. I use brown thread on the inlays. This prevents ruining the inlay fabric if you decide to antique or color the thread. Starting at the center point of the lobes, work your way around and end with two back stitches. Repeat the process for the two circles flanking the heart.

Figure 47. This photo illustrates the back side of the flap, with the inlays stitched through both flap pieces. If you prefer, you can stitch the inlays to the top piece only, but this must be done before the two flap pieces are sewn together.

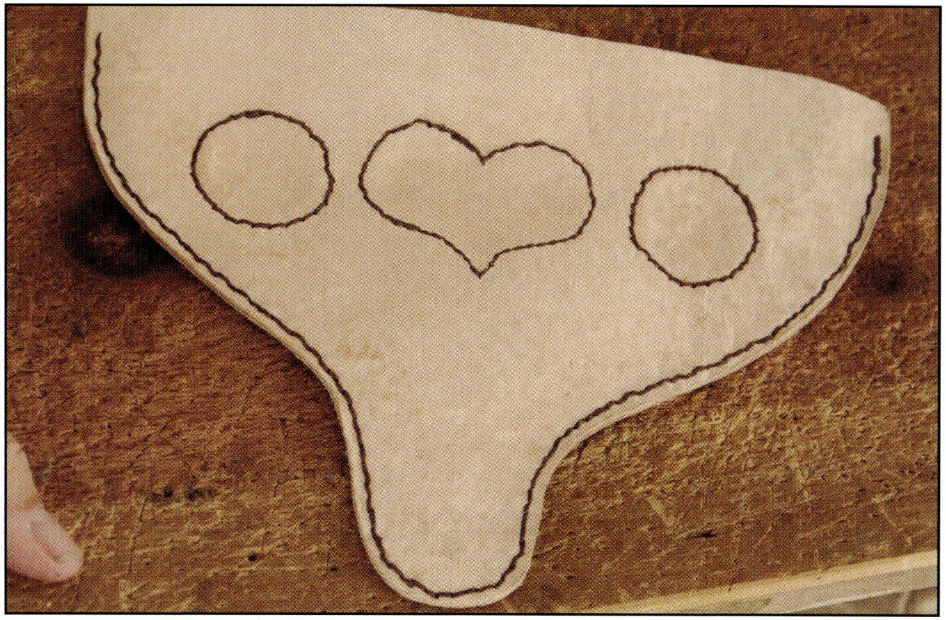

47. The back of the stitched flap.

48. Trim the flap edge with a sharp knife.

49. Burnish the flap edge with a wooden handle.

50. The burnished flap edge.

Figure 48. Trim any uneven edge of the flap with a sharp knife or shears. Be careful not to cut the stitching.

Figure 49. Dampen the edge with water using your finger. Burnish the entire edge using a wooden handle, such as the one on your awl, smoothing the roughness and imparting a near seamless finish. You will end up with a nice looking burnished edge as shown in **Figure 50.**

Old leather cowboy boots are a good source of leather for making small pouches. This pouch measures 7" wide, by 6-1/2" long.

Constructing the body:

51. Line up the body pieces with the rough sides out.

52. Dampen the edges of the two body pieces.

Figure 51. Align the front and back body panels, placing them smooth side to smooth side. In a later step, after you stitch the body together, you will flip the body inside-out so that the smooth sides are facing outwards.

Figure 52. Dampen the edges of the front and back body pieces with warm water using your finger. The shorter body panel will be the front of the pouch.

Figure 53. On the rough side of the short body panel, measure $^3/_8$" down from top, and $^3/_{16}$" from the edge. Make a mark on the left corner. Make the same measured mark on the opposite side. You will lay out your stitches around the body between these marks.

53. Measure the edge for the stitching.

54. Lay out the stitches using an over-stitch wheel.

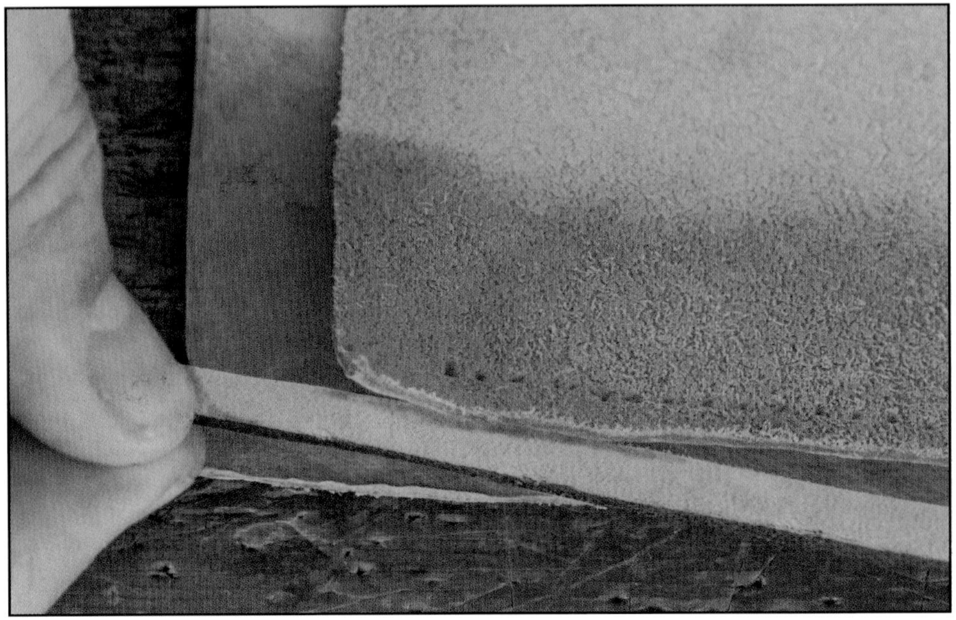

55. Insert the leather welt between the body.

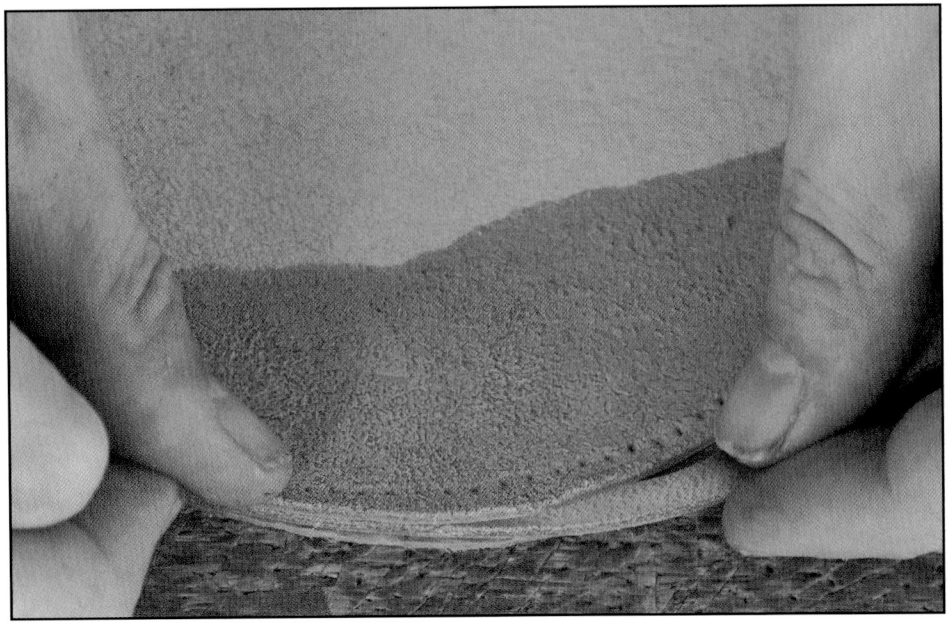

56. Bend the welt around the curved edges of the body.

Figure 54. Using an over-stitch wheel, carefully lay out the stitch marks. You are marking the rough side of the leather, so they will be a little less distinct. Carefully roll back and forth a few times and really press the marks in distinctly.

Figure 55. Fit the welt between the front and back panel. The welt is a long $^3/_8$" wide strip of leather sewn between the body pieces. When fitting the welt strip, you must be sure to have the smooth side of the welt facing the smooth side of the front panel. This is important when you turn the bag inside out and the welt shows against the panels. Do not be tempted to glue the welt in place. Glue will really stand out and look unpleasant when you dye the leather.

Figure 56. Dampen the welt and fit it completely around the body. You may note that the original North Hampton County bag on page 12 has a gusseted panel with welted seams sewn into the bottom. The gusset has been omitted here to simplify this project. The more difficult gusset will be discussed and illustrated later in this book.

57. Punch keeper stitch holes using an awl.

58. Temporarily tie the three pieces together.

Figure 57-58. Align and punch a keeper stitch hole through all three pieces. Thread and tie each keeper stitch before you continue to the next, working your way around the body, ending at the opposite corner. I added seven keeper stitches along the marks laid out by the over-stitch wheel. These keep the welt from shifting while punching awl holes and stitching the body together.

Figure 59. Punch the remaining holes through all three layers of leather using an awl. Keep the awl holes aligned and centered on the welt as you punch around the edge of the body. You should be able to feel the inside edge of the welt through the body with your fingers. This will help you keep the holes centered. Saddle stitch the top corner. Start with a few back stitches to reinforce the stress point at the corner of the pouch. Keep each stitch in the center of the welt strip and the stitches running uniformly close to the edge.

Stitch around the body, removing the keeper stitches as you go. If you run short on thread, back stitch before you run out and trim flush. Thread your needles, and start stitching again where you left off. Do not leave a gap between the old and new stitches. End the stitching with at least two back stitches, to reinforce the stress point at the corner of the pouch.

59. Punch the holes using an awl. Saddle stitch the body together.

60. Trim the edge flush with a knife or shears.

61. Burnish the edge with a wood handle.

62. The body, stitched together and trimmed.

Figure 60. Trim away any uneven edge of the body with a razor knife or shears. Do not cut your stitches.

Figure 61. Dampen the edge with warm water, and rub with the handle of your awl to burnish the edge. The body should look like **Figure 62**. The body is now ready to be turned inside-out, so the smooth side is facing outwards.

Figure 63. Soak the bag in warm water. A good soaking is needed to soften the leather, making it easier to flip inside-out. If the stitches are tight enough, the bag should hold water like a bucket. You can be proud if it does.

Figure 64. Flip the body inside-out. A little pushing and pulling will be needed. Pay close attention to the seams making sure the welt is uniformly exposed along the edge. Puckers, folds, or wrinkles should be worked out as you go. Re-soak the body if needed. You should not have much trouble turning a pouch made of 3-4 oz. leather.

63. Soak the body in water to soften.

64. Flip the body inside out.

Figure 65. This view shows a close-up of the raw edge of the welt. If needed, trim any uneven edge of the welt with a sharp knife or shears. Be very careful not to cut the body!

An alternative method of welting a seam is to fold a thinner piece of leather in half, and sew it between the body so that the folded edge will be exposed when reversed as shown below. This method is a little more difficult, but has an attractive finished appearance.

65. The finished welted seam.

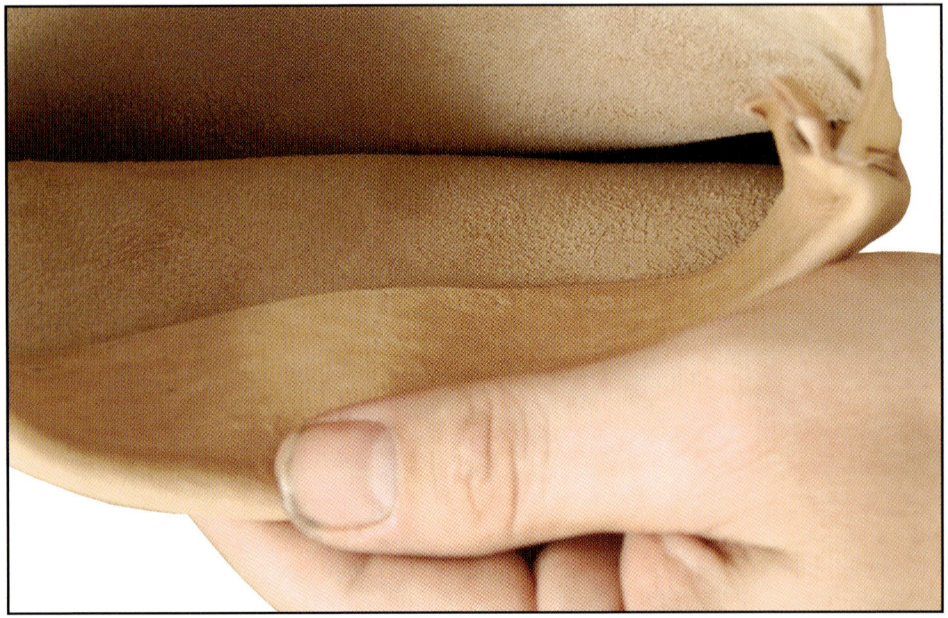

66. Fold the lip of the front panel inside the pouch.

67. The front side of the body with folded top edge.

68. The folded lip and the corner reinforcement piece.

Figure 66-67. Dampen the front lip of the pouch. Fold $1/4$" to $1/2$" of the lip of the front panel to the inside of the pouch making a hem. This step is not necessary, but it will contour and stiffen the lip to allow easier access for your hand.

An alternative way to finish the opening is to cut a round slot as shown below. This offers more room for your hand when reaching into the bag.

69. Fold the reinforcement piece over the top corner.

70. Lay out the stitches with an over-stitch wheel

Figure 68. Find your corner reinforcement with decorative edge. Notice the clipped corner which will face toward the front of the pouch.

Figure 69. Soak the corner reinforcement piece in water and wrap it around the top edge of the bag. Fold the reinforcements over the upper corners with the clipped corner toward the front of the pouch. These pieces lessen strain on the corners of the pouch during use.

Figure 70. Lay out the stitches with an over-stitch wheel. Start the over-stitch wheel at the base of the clipped corner, and run the stitches just past the folded hem of the front panel to capture the hem.

Figure 71. Keep the reinforcement tight against the side of the pouch body. Punch the over-stitch holes with an awl. You must go through the back edge of the reinforcement piece. Do not wander off, or make a crooked run. Stitch the reinforcement with saddle stitches, finish with at least two back stitches trimmed flush. Attach the opposite reinforcement repeating the same steps.

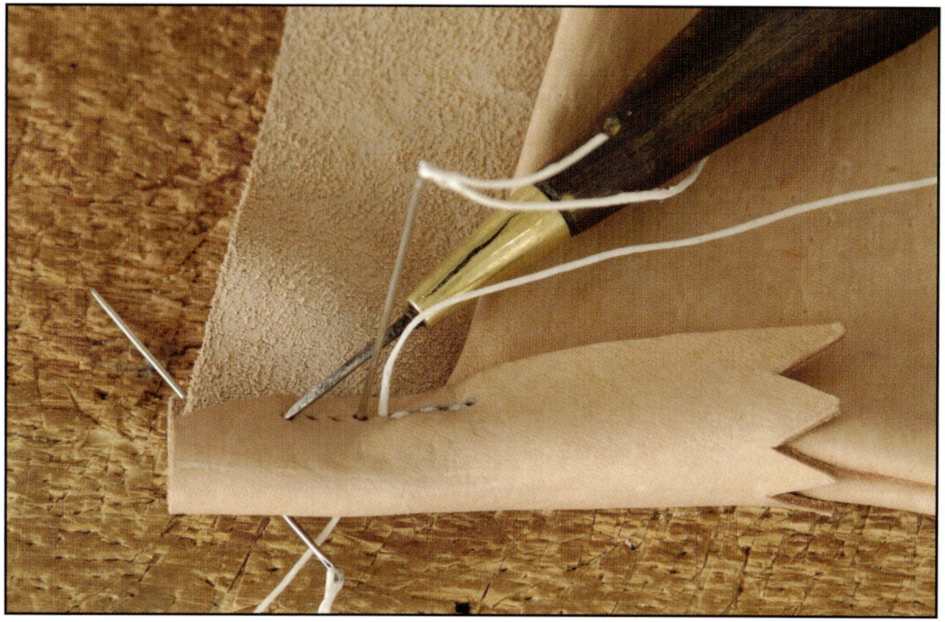

71. Saddle stitch the leather reinforcements to the body.

Construct the buckle shoulder strap:

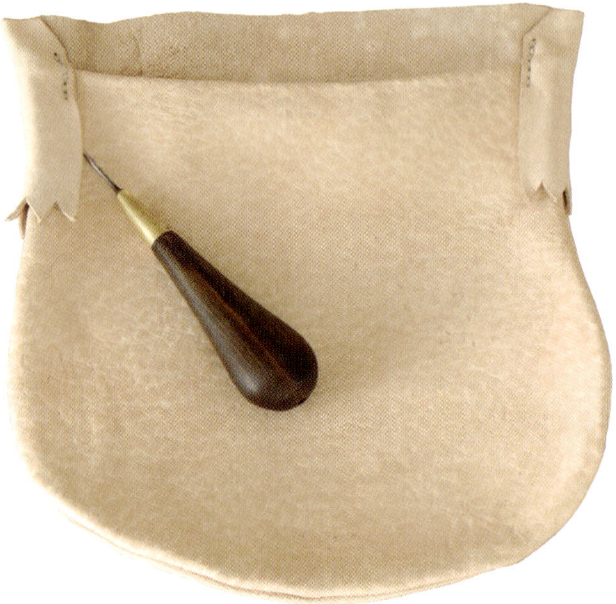

72. The finished body.

73. Measure the length of the buckle shoulder strap.

74. Trace the bottom of the buckle strap pattern onto the leather.

Figure 72. The finished body reinforcements saddle stitched in place.

Figure 73. The shoulder strap will be made of two pieces, a shorter buckle strap and a longer adjustment strap. Consider two things before making the straps: do you want the buckle on the front or back when the pouch is worn? Do you wear your pouch on the left or right side? The basic assembly is the same for both. The pouch shown is constructed to be worn on the right with the buckle on the back side.

Cut a strip of leather 2" wide and a minimum of 20" long. Or measure the distance from the top of your pants back pocket to the center of your back, below your shoulder blade. Add up to two extra inches.

Figure 74. Use the template provided in the North Hampton County Pouch pattern labeled "Buckle Strap". Trace the pattern end marked "Buckle Strap Pouch End" (with the offset edge) on the smooth side of the leather. Cut with shears or a razor knife. The indent will be located on the outside of the strap when attached to the left side of the pouch as you face it. See page 50 for a finished example.

75. Trace the buckle strap pattern onto the leather.

76. Check your lines and brass buckle before cutting.

Figure 75. Trace the pattern "Buckle Strap Buckle End" on the opposite end of the leather strap with a soft pencil, making a 3" long tongue.

Figure 76. Check the distance between the inside edges of your buckle against the lines you have drawn before you cut.

Figure 77. Cut the strap with a razor knife or shears. The leather strap should fit snugly in the buckle

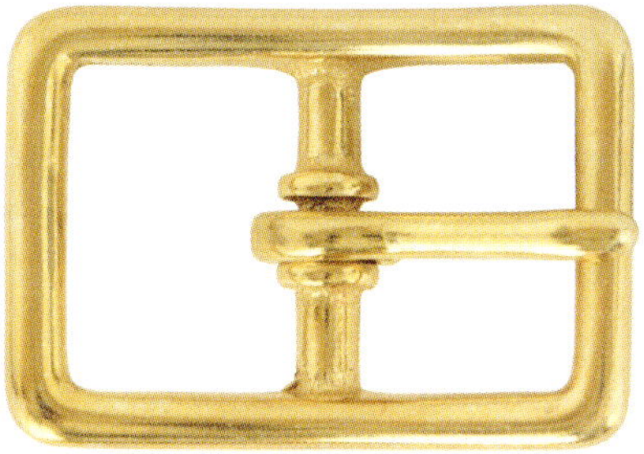

For this project I used a cast brass buckle with brass tongue. Shown actual size.

77. Cut the strap end with a razor knife or shears.

78. Make sure the strap fits the buckle snugly.

79. Fold the buckle strap.

80. Cut a hole for the buckle tongue with a knife.

Figure 78. Test fit the buckle. The strap should fit snugly. Mark a spot for the buckle tongue with a pencil or awl 2” from the end of the strap.

Figure 79-80. Fold the leather over at the two inch mark. Using a razor knife cut a $^3/_8$” deep “V” in the center of the creased fold.

The original buckle.

81. The finished "V" cut for the buckle tongue.

82. Dampen the buckle strap to soften.

Figure 81. When unfolded, you will have a nice little diamond that will mold around the tongue of the buckle.

Figure 82-83. Dampen the leather. Insert your buckle tongue through the "V" cut. The folded leather should be rough side to rough side. Using an over-stitch wheel, lay out at least ten stitches from the end of the strap to the buckle center bar. Repeat on the opposite edge. You can connect these two rows across the square edge with either the over-stitch wheel or by eye.

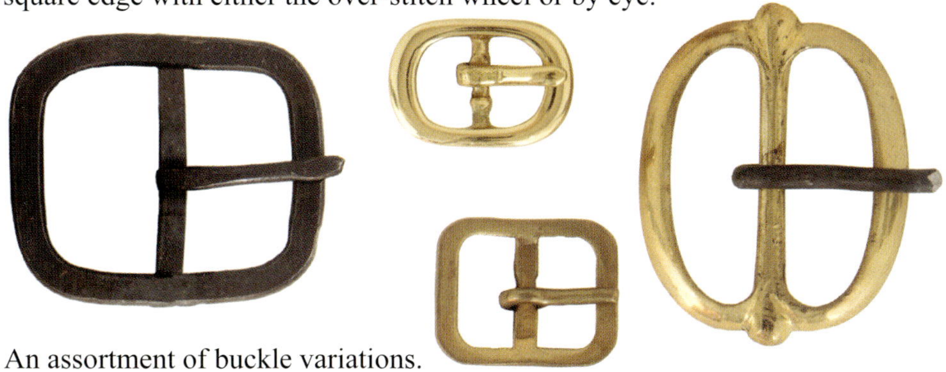

An assortment of buckle variations.

83. Lay out the stitches with an over-stitch wheel.

84. Punch the stitching holes with an awl.

85. Saddle stitch the buckle strap together.

86. Cut slots for the strap keeper.

Figure 84-85. Punch the stitch holes using an awl. Begin the saddle stitch at the cross bar of the buckle. Finish with at least two back stitches, securing the buckle to the strap.

Figure 86. Cut a 1/2" wide strip of leather, about 4" long for a strap keeper. The strap keeper will be mounted about 3" from the center bar of the buckle on the smooth side of the strap. At the 3" mark, 1/4" from the edge, cut a slot parallel to the strap with a 1/2" wide buttonhole chisel. If you do not have a buttonhole chisel, a 1/2" wood chisel, or a razor knife will work. Do not make the slot any wider than the keeper strap.

Figure 87. Slide the strap keeper or "retaining loop," smooth side up, through the cut slot.

Figure 88. On the opposite side of the strap, cut a second parallel slot 1/4" in from the edge with the 1/2" buttonhole chisel.

87. Slide the strap keeper through the slot.

88. Cut the second slot with a buttonhole chisel or knife.

Figure 89. Insert the strap keeper into the second slot. If necessary you can dampen the strap to make sliding the strap keeper in the slot easier. The strap keeper should remain dry.

Powder was seldom poured directly from a powder horn into the barrel of a rifle or fowler. Instead, the proper amount of powder was measured using a metal, horn, bone, or antler powder measure, such as this one made of deer antler. Powder measures were commonly hung from the shoulder strap of the hunting pouch using a leather cord.

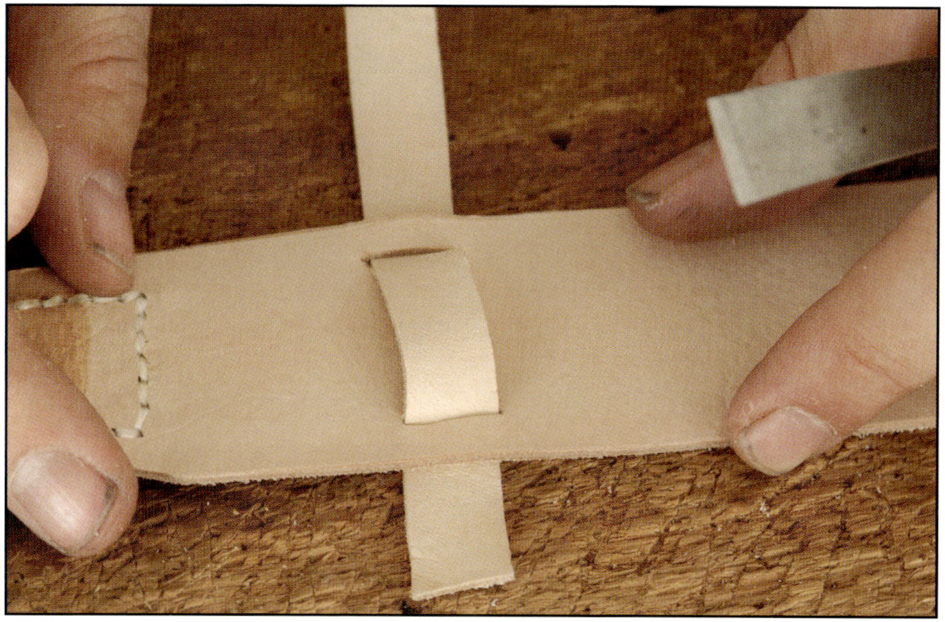

89. Slide the strap keeper through the second slot.

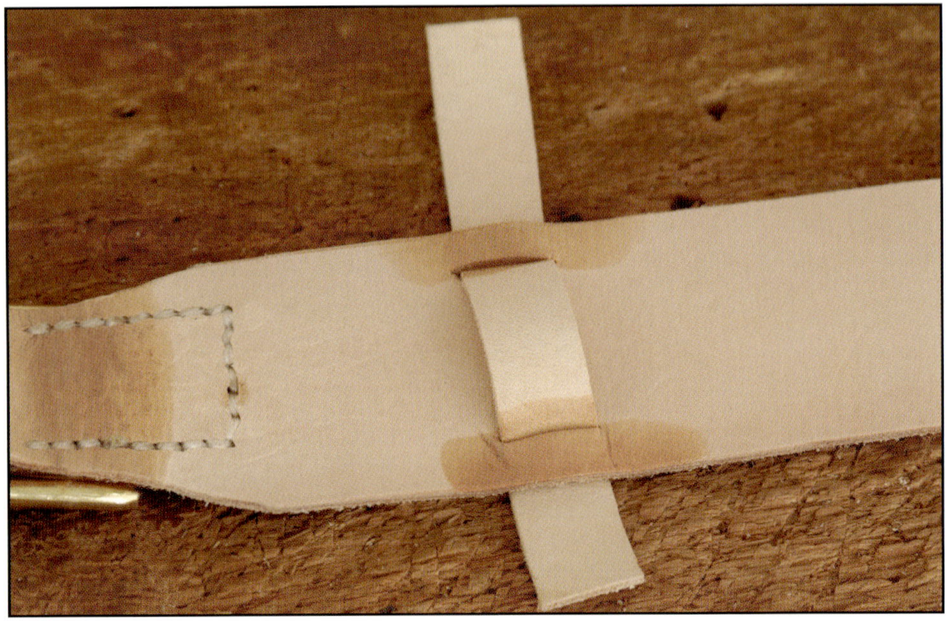

90. Dampen the edges with water to soften.

91. Lay out the stitches with an over-stitch wheel.

Figure 90-91. Dampen the leather and mark out the stitches using an over-stitch wheel, with one stitch on either side of the strap keeper. Use your thumbnail to delineate the ridge the strap keeper makes against the strap. This will help you to tightly make the first and last stitch off the edge of that ridge.

Figure 92. Punch the stitch holes with an awl. At seven stitches per inch, there should be six stitches.

92. Punch the stitch holes with an awl.

Reproduction carved bone powder measures.
Track of the Wolf archives.
www.trackofthewolf.com

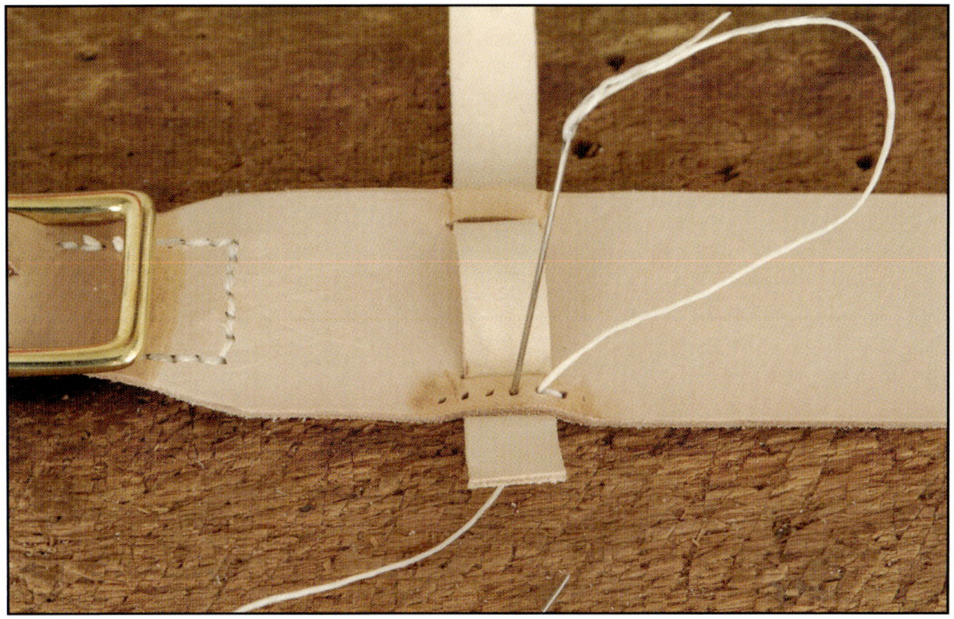

93. Saddle stitch the strap keeper to the shoulder strap.

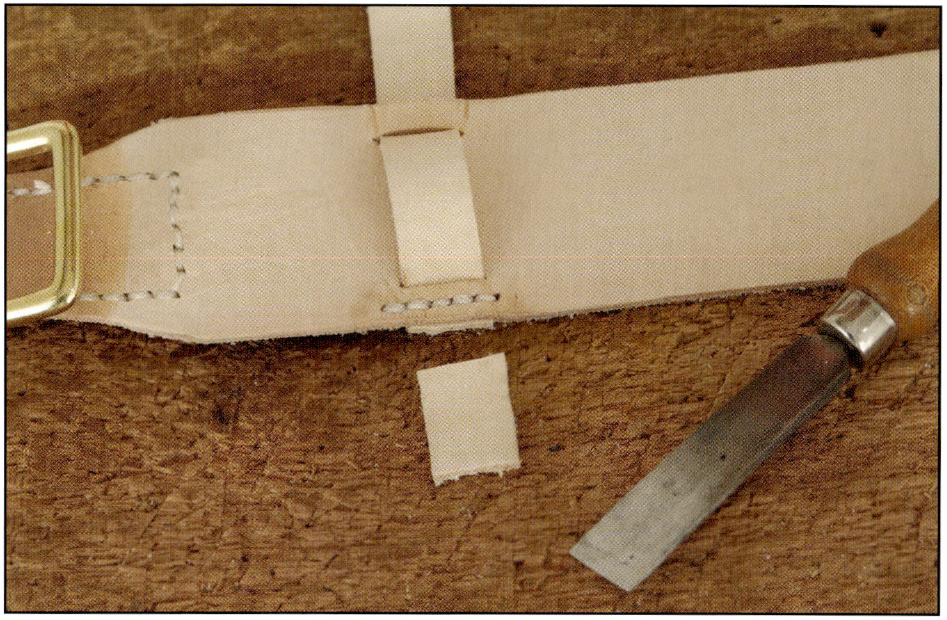

94. Cut the loop ends flush with the shoulder strap edge.

Figure 93. Thread two needles and saddle stitch one side of the strap keeper. You may want to back stitch all of the stitches, except the last one.

Figure 94. On the side you stitched, trim off the excess strap keeper flush with the edge of your strap. The buttonhole chisel works great, but a sharp knife also works. Repeat the steps for the opposite edge. When done, the buckle strap should look like the one shown in **Figure 95**.

Original carved bone and brass powder measures.
A. Rehder collection.

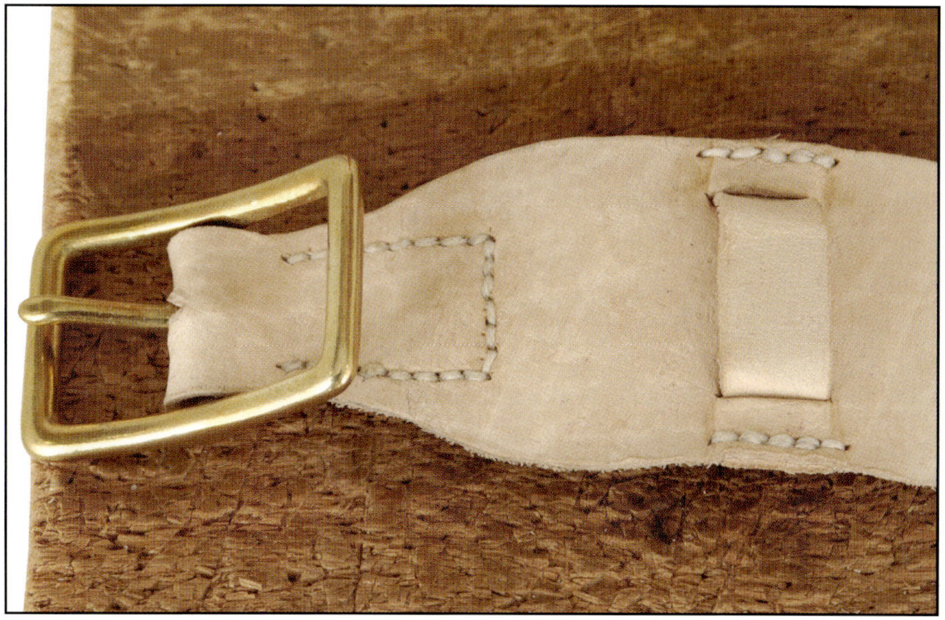

95. The finished buckle strap.

Construct the adjustment shoulder strap:

96. Trace the bottom of the adjustment shoulder strap.

97. The bottoms of both shoulder straps.

98. Measure 12" from the opposite end of adjustment shoulder strap.

Figure 96. The longer adjustment shoulder strap is made with a few simple cuts. An average size strap measures about 53". This should correspond to the measurement from the top of your right back pants pocket, up and over your left shoulder, down the center of your chest to just below your navel. If there's any doubt, make it longer. It is always easier to trim the strap shorter than to splice it longer. Cut a 2" wide strap 53" or longer.

Figure 96-97. Using the North Hampton County pouch pattern template labeled "Adjustment Strap Pouch End", mark and cut the strap with indent. It should be a mirror image of the buckle strap.

Figure 98. On the opposite end, trim a 1" wide tongue that will fit in the buckle. The 1" wide tongue should be at least 12" long and then taper gradually to the full 2" strap width. Use the pattern provided to help you draw the taper to match the other strap.

99. Extend the pattern line from the base of the pattern to the end of the strap.

100. Cut the 12" long tongue with a razor knife or shears.

Figure 99. Extend the pencil lines creating a 1" wide tongue on a 2" wide strap. Remove 1/2" from each edge of the strap, from the tip, to the where it flares to a full 2" width. Use a marking gauge, or simply mark it freehand by using the edge of your forefinger running along the edge of the strap as a fence.

Figure 100. Following the lines laid out, cut the 12" tongue with a sharp razor knife or shears.

Figure 101. Draw in a rounded tip for the adjustment strap.

101. Using a pencil, draw a rounded end on the tongue.

102. Cut a rounded tip on the adjustment strap tongue.

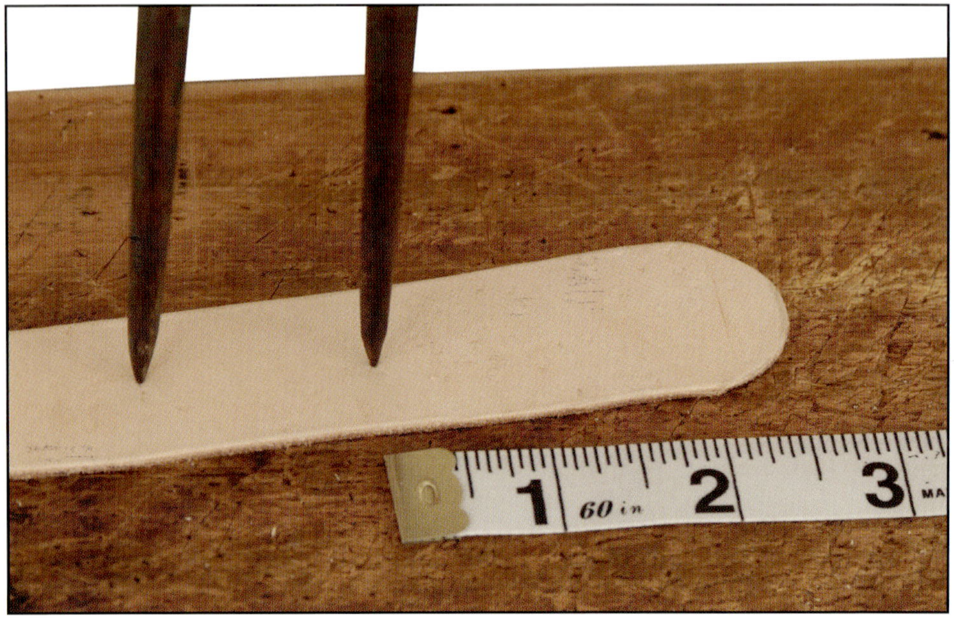

103. Measure the buckle adjustment holes.

104. Mark six or more adjustment holes at regular intervals.

Figure 102. Cut a rounded tip on the adjustment strap tongue with a sharp razor knife or shears.

Figure 103-104. Set a pair of wing dividers to 1-$^1/_2$". Measure 2" from the rounded tip and make the first mark. Walk the wing divider to make another mark 1-$^1/_2$" from the first, and again for the second. Make six, or seven evenly spaced marks. Now you have the holes for your buckle marked.

Figure 105. Compare the buckle tongue mounted on the buckle strap against your rotary punch sizes, and select the proper size punch.

A fancy adjustable powder measure with a turned head.

105. Use the buckle tongue to select a correct diameter hole punch.

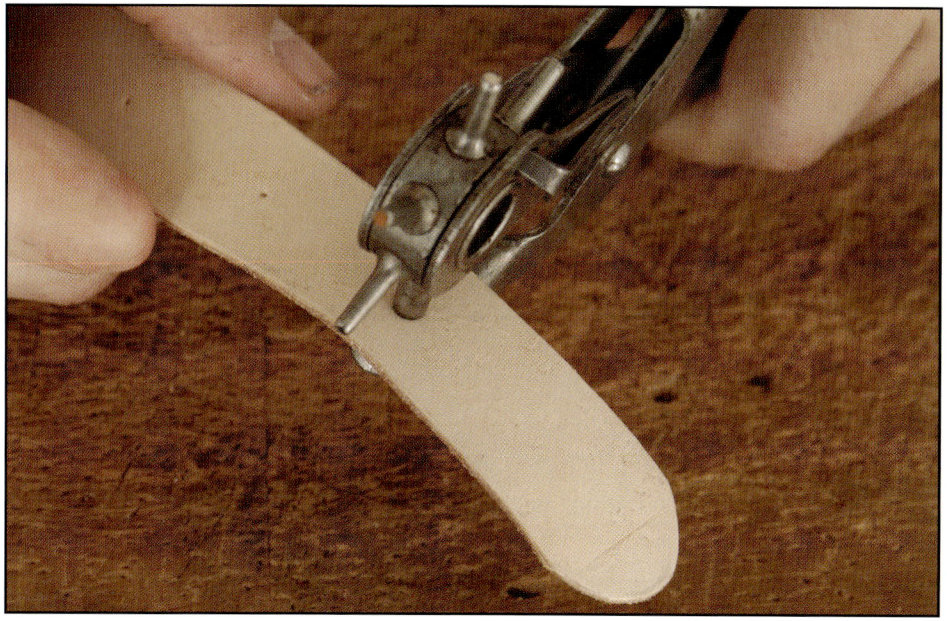

106. Punch the strap adjustment holes.

Figure 106-107. Using the punch size you have selected. Punch all of the measured marks located on the adjustment strap. Dampening the leather allows you to punch holes more easily than when dry. You now have a completed set of straps, and are ready to make the powder horn keepers.

An original small leather pouch
sewn with leather lace,
circa 1850-1900.
Track of the Wolf archives.
www.trackofthewolf.com

107. The adjustment strap cut, with holes punched.

Attach the powder horn straps:

108. Measure the strap width using the buckle, and mark with a pencil or awl.

109. Cut a pair of powder horn straps with a knife or shears.

110. Fold the strap, and cut out a notch for the buckle tongue with a knife.

Figure 108. Take one of the powder horn strap buckles and measure the inside width. I am using 1/2" buckles. Transfer the width to the leather by marking it with a soft pencil, or awl.

Figure 109. Cut a pair of short straps, with your shears, about 3-3/4" long, by 1/2" wide, or as wide as the buckle you have selected.

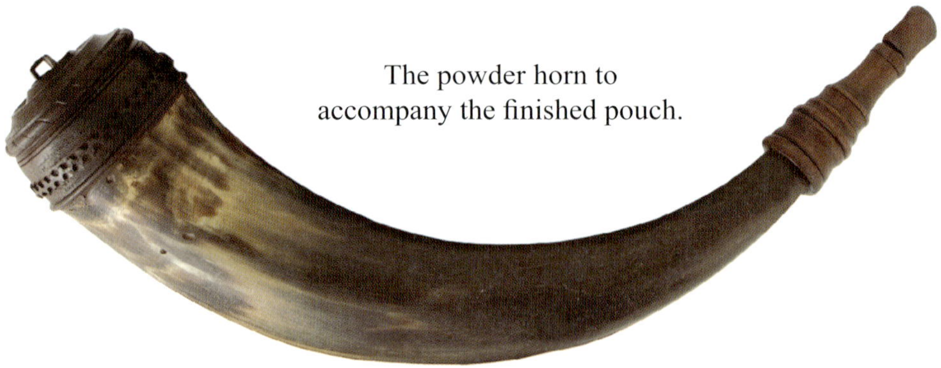

The powder horn to accompany the finished pouch.

111. Skive the tips of straps.

112. Trim off any excess leather that appears ragged.

Figure 110. Fold the powder horn keeper strap, rough side to rough side, about 1-1/2" from the end. The top of the strap should be about 3/4" longer than the bottom, so that one layer, not two, will be sewn onto the shoulder straps. Cut an appropriate size "V" notch for the buckle tongue with a sharp knife. Repeat on the second powder horn keeper strap.

Figure 111. Skive the rough side of both ends of the straps about 3/4". The long end will lay on the front of the shoulder strap. The other end will be sewn behind the buckle. Skiving will make a nice looking powder horn keeper strap.

Figure 112. Trim off excess leather that may appear too thin or ragged.

Figure 113. Dip the powder horn keeper straps in warm water to soften for folding them around the horn keeper buckle.

113. Dampen the powder horn keeper straps to soften.

114. Insert the buckle and lay out the stitches using an over-stitch wheel.

115. Punch the stitch holes with an awl.

116. Stitch the two layers together, stopping at the single layer.

Figure 114. Insert the buckle tongue in the cut "V" notch. The tongue must point down like the buckle. Starting near the buckle, about $^1/_8$" from the edge, lay out the stitches using an over-stitch wheel. When finished, repeat on the other side.

Figure 115-116. Punch the stitch holes with an awl. Begin your saddle stitching at the buckle cross bar. Once you have stitched the two layers of leather together, and reach the single layer, stop, and position the powder horn keeper strap on the shoulder strap.

Figure 117. Position the pair of powder horn buckle straps centered about 8-$^1/_2$" from the ends of both shoulder straps. Use the existing awl holes on the powder horn buckle strap as a guide. Punch awl holes through the shoulder straps with an awl.

Figure 118. Saddle stitch the remaining stitch holes through the powder horn buckle strap and the shoulder strap. Remember, the buckle points down towards the pouch end of your shoulder straps.

117. Position the horn strap. Punch awl holes through the single layer only.

118. Saddle stitch the single layer of the buckle strap to the shoulder strap.

Figure 119-120. Saddle stitch the single layer of the powder horn buckle strap to the shoulder strap, finish stitching the two layers of the powder horn buckle strap together up to the buckle. The powder horn keeper strap should hang free except for the last five stitches on either side. Do not get carried away and stitch it totally down to the shoulder strap!

An original buffalo powder horn
courtesy of A. Rehder

119. Stitch the remainder of the powder horn strap.

120. The powder horn keeper strap attached to the buckle strap.

121. Repeat the process on the adjustment shoulder strap.

122. Measure the length of the powder horn adjustment straps.

Figure 121. Repeat the process for the powder horn keeper strap on the adjustment shoulder strap. Next you need to make a pair of adjustment straps to attach to the powder horn.

Figure 122. Use a long piece of strap leather, or cut it from the edge of a larger piece. Mark a piece of leather 16" long, and $^1/_2$" wide, or as wide as your selected buckle. You will need two horn adjustment straps.

A small flat priming horn, a handy accoutrement to keep in your pouch.
Made by author.

123. Use a buckle to mark the width of the horn adjustment strap.

124. Cut out the powder horn adjustment straps with a knife or shears.

Figure 123-124. If your are cutting the two horn adjustment straps from a larger piece of leather, mark the inside width of the buckle on the leather with a pencil or an awl. Draw a straight line across the leather. Cut the strap with a sharp razor knife or shears.

Figure 125. Set a wing divider to 3/4". Starting 1-1/2" from one end of the horn adjustment strap, make a mark. Walk the wing divider to make additional marks until you have marked four buckle holes. Select the proper punch size on the rotary punch to match the buckle tongue. Punch four holes in each strap. Trim the end near the holes to a blunted point. Your pair of horn adjustment straps should resemble the one shown. Wetting and stretching the straps will elongate the holes giving them a nicer shape, rather than being perfectly round.

125. Punch the adjustment holes with a hole punch.

Attach the shoulder straps to the body:

126. Lay out the stitches on the shoulder straps with an over-stitch wheel.

127. Line up the strap and body. Punch the holes through both with an awl.

128. Saddle stitch the strap to the body.

Figure 126. Mark a row of stitches across the bottom of one of the strap pieces with an over-stitch wheel, at least $1/8$" from the edge.

Figure 127. You have the option to attach the straps straight, or angle them outwards. I will install an angled strap on this pouch. An angled strap conforms better, and reduces stress on the stitching. An angle of 20 degrees is sufficient. Position the strap close to the reinforced edge on the corresponding inside rear panel of the pouch. The end of the strap should overlap the top of the body $3/4$". Keep the angled bottom of the strap parallel to the top edge of the pouch. Punch the stitch holes through the strap and rear pouch panel using an awl.

Figure 128-129. Saddle stitch the strap to the pouch. Back stitch two stitches, finishing with both needles on the back of the pouch. Trim the threads flush with a razor knife.

Figure 130-131. Repeat the process for the opposite shoulder strap. Notice that the straps angle up and away from the pouch. This will help contour the pouch around your torso.

129. Trim the excess thread with a razor knife.

130. The buckle strap sewn to the body.

Pan brushes and vent picks were sometimes attached to the shoulder straps of hunting pouches for easy access. Some were attached with simple leather straps, other were attached with iron, or brass chains. This "whisk and pick" has a simple wire vent pick, and a straw pan brush wrapped with brass wire.

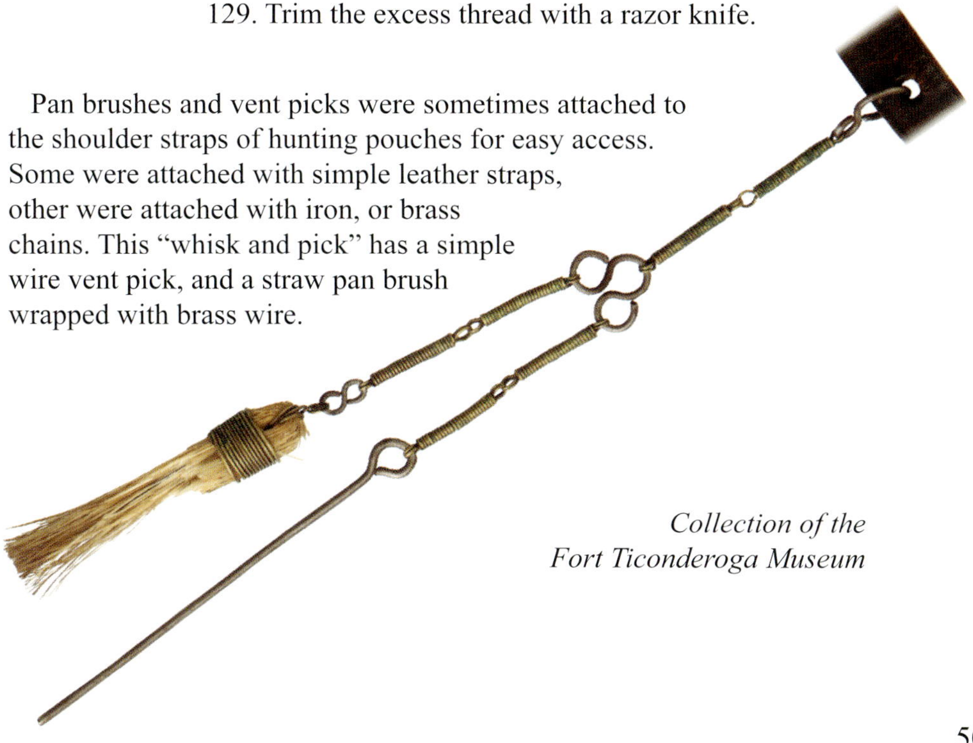

Collection of the Fort Ticonderoga Museum

131. Repeat the process on the adjustment strap.

Attach the flap to the body:

132. Dampen the top edge of the pouch flap.

133. Lay out the stitches with an over-stitch wheel.

134. Line up the flap and the body.

Figure 132. Dampen the top edge of the pouch flap. The over-stitch wheel marks will be easier to see.

Figure 133-134. Lay out the stitches with an over-stitch wheel across the top of the pouch flap, about 1/8" from the top edge. Center the flap on the pouch body. The flap may extend beyond the edge of the body, which is okay. Make sure any overhang is equal on each side. The pouch flap has about three stitches of overhang. If your pouch flap is slightly wider or shorter, that is okay. You can stretch the pouch flap by wetting the top edge and pulling it from opposite edges. Once stretched, you can not shrink it.

Figure 135. Punch the stitch holes with an awl, starting from one end and working toward the other. Your awl will punch holes through up to five layers of leather at one time. The awl holes must be straight on the front and the back of the pouch. Punch the awl through all leather layers consistantly, straight up and down, perpendicular to the pouch.

135. Using an awl, punch stitch holes through the flap, body, and straps.

136. Temporarily tie the flap to the body with thread.

Figure 136. Tie four or more keeper stitches, one on each side against the straps, two equidistant across the remainder of the pouch flap edge. This will ensure that the flap stays centered as you attach it.

Figure 137. Saddle stitch the flap to the body. The first three or more stitches are only through the overhanging flap. The next stitch will go through the flap and the body at the reinforcement. You will be stitching through five layers of leather. These stitches need to be tight. Double stitch for additional strength.

137. Saddle stitch the flap to the body, through all of the layers.

A simple powder horn with carved head,
circa 1820-1890.
Track of the Wolf archives.
www.trackofthewolf.com

138. Stitch the pouch flap to the body.

139. Stitch through the shoulder strap.

Figure 138-139. Saddle stitch across the top of the pouch flap, capturing the shoulder straps as you go.

Figure 140. Slip the powder horn adjustment straps through the buckles. When you have finished, your pouch should look like the example. Take a minute and try on your pouch to test the fit.

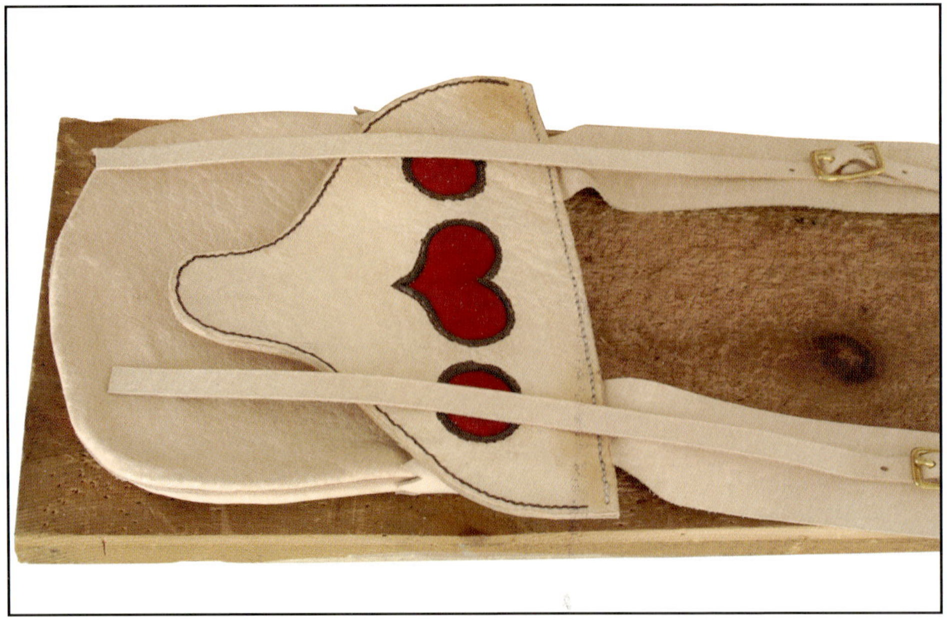

140. Attach the powder horn straps. Try on your pouch.

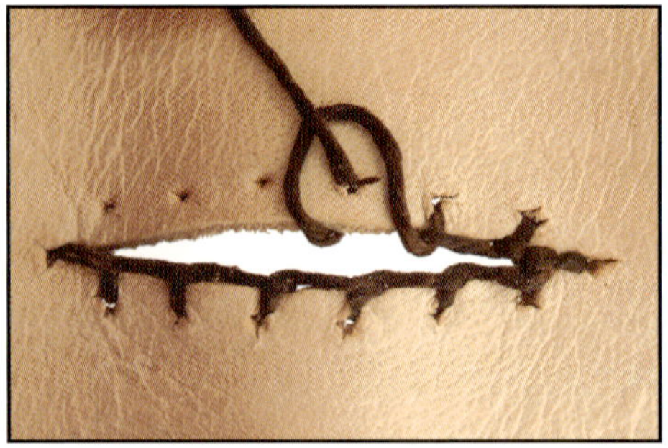

Some flaps close with a button and buttonhole. Use a buttonhole stitch to reinforce the edges. Punch a row of holes along each edge of the slit. A traditional buttonhole stitch comes up from the bottom, and threads the needle through the previous stitch, forming a slip knot. Pull the thread tight so the knot rests along the raw edge of the buttonhole.

Powder horn
keeper strap
Powder horn
adjustment strap

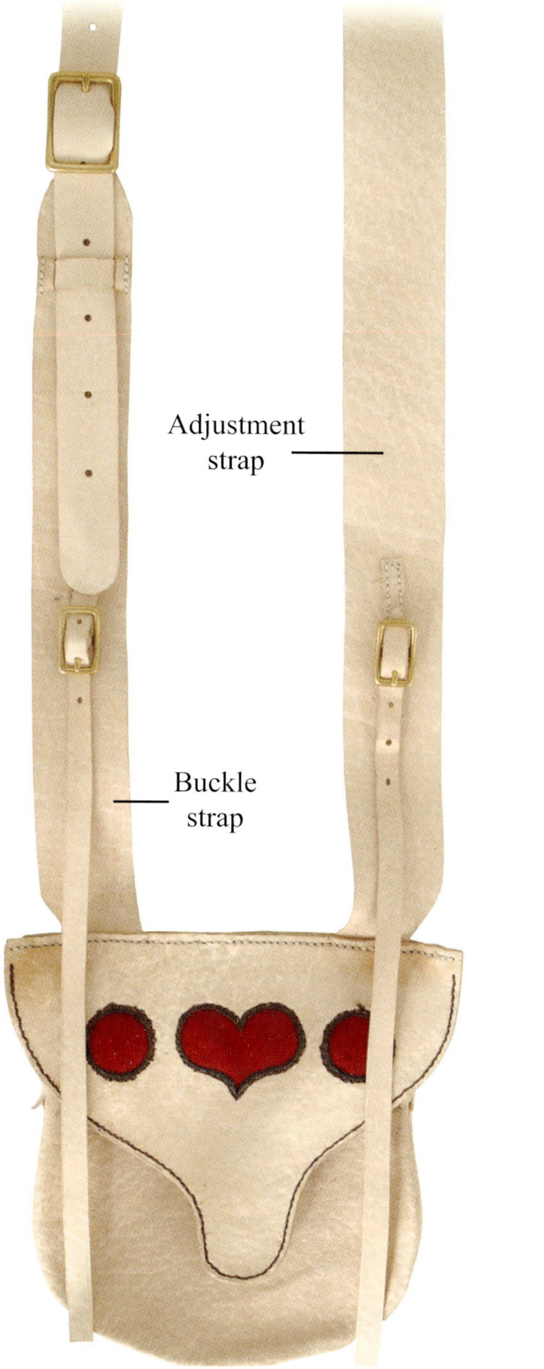
Adjustment
strap
Buckle
strap

Making decorative edges:

141. Trace around a coin as a pattern, for a decorative edge.

142. Cut relief at the intersections with a knife or buttonhole chisel.

Figure 141. To add a decorative edge or welt to your pouch, use a coin the size you want your scallops. Draw a straight line across your leather about half the coin diameter. Trace around half of the coin holding it along the edge of your leather. Walk the coin and continue tracing scallops, making a strip as long as you require.

143. Cut the scallops with a knife of shears.

An antique brass powder flask.
Track of the Wolf archive.
www.trackofthewolf.com

144. Punch decorative holes with a rotary punch.

145. Cut off the excess leather, leaving enough edge to stitch into the pouch.

Figure 142-143. Make a relief cut at the points where your scallops meet, do not cut any deeper than your center line. A buttonhole chisel works well. Use shears to carefully cut the scallops.

Figure 144. For another decorative touch, punch a hole in the center of each scallop using a rotary punch.

Figure 145. Trim the excess leather, leave $3/8$" above the pencil line to stitch to the body of the pouch.

Figure 146. The scallops can be applied as a decorative welt between the body pieces. If the scalloped welt is stitched between pieces that will be turned smooth side out, the scallops must be on the inside between the two pieces, otherwise when you turn your pouch right side out, you will discover your scallops are neatly decorating the inside of your pouch where no one will see them. If the pouch is to be stitched together and not reversed, stitch them as shown.

146. Stitch the decorative welt between the body pieces.

Dyeing the pouch:

147. Prepare the dye.

148. Apply the dye around the cut out using a small brush.

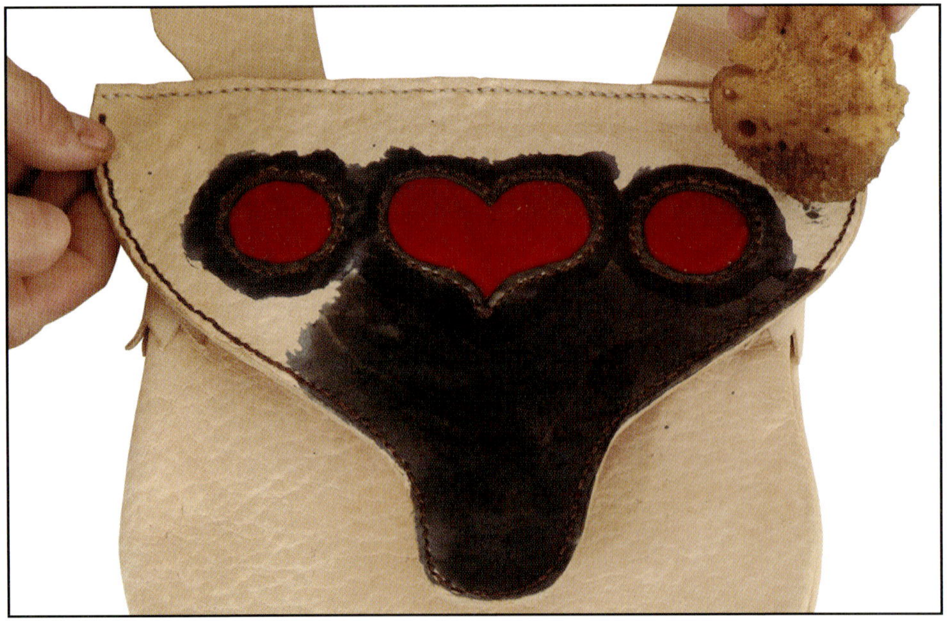

149. Dye the flap with a sponge.

Figure 147. Before applying any dye, your pouch must be dry. I will continue using the home made harness dye made from vinegar and rusted iron, which I used to dye around the inlays. The process for making the harness dye is explained in detail on page 19. You can also use commercial leather dye such as "Feibings" sold by leather supply stores. These dyes come in a variety of colors. Brown or black are the best choices. If you are not using vegetable tanned leather, you should use a commercial dye.

When working with dye, wear rubber gloves, and use a plastic drop cloth. Adequate ventilation is needed, especially if you cannot work outside.

Figure 148-149. Use your preferred leather dye, a small brush, and a piece of sponge. Apply the liquid dye to the dry pouch. Start with the flap. Use the small brush to apply around the inlays, being careful not to wet the inlay fabric. On large areas you can use a small sponge. The nice thing about the home made harness dye is that it only reacts with the tannin in vegetable tan leather. Generally it will not transfer color to your skin or surroundings if spilled.

150. Dye the edges of the flap.

151. Dye behind the corner reinforcement.

Figure 150. Continue to dye the flap. Apply generous amounts of dye to all of the edges.

Figure 151-152. Dyeing the back panel is straightforward. Apply dye behind the reinforcements and cover any streaks with repeated applications.

A strike-a-light is an essential tool in every hunting pouch.

152. Dye the back of the pouch.

153. Dye the edges of the pouch.

154. Dye the straps.

155. The completely dyed pouch.

Figure 153. Continue applying dye. Apply generous amounts of dye to the seam and welt. Don't forget to dye the area underneath the corner reinforcements. Finish the inside of the flap and the visible portions of the pouch interior.

Figure 154-155. Dye the straps, front, back, and edges. The iron based harness dye works on contact. It dyes even the rough side of the vegetable tanned leather very easily.

Figure 156. Allow the pouch to dry. Apply a colored wax polish on the smooth exterior surfaces. Use a brown shoe polish applied with a dauber or toothbrush. Pay special attention to all of the visible stitches. The brown polish will color white threads, blending them nicely with the pouch. Do not allow polish to touch the inlay fabric.

156. Darken the stained leather with shoe polish, or shoe black.

157. Rub off the excess shoe polish with a dry brush.

Figure 157. Using a dry boot brush, polish the waxed surfaces as if you were polishing a pair of shoes. This will impart a nice finish to the pouch.

Figure 158. Burnish the exposed edges with the handle of your awl.

Trade silver was sometimes used to decorate the flaps and shoulder straps of old pouches.

158. Burnish the edges with a wooden handle.

Making a knife sheath:

159. Trace the pattern onto leather with a pencil.

160. Cut the sheath with a knife or shears.

161. Add a "pinked" or serrated edge.

Figure 159. If you have a patch knife, you may wish to make a sheath, and attach it to the pouch. The pattern included is a basic pattern that will hold a knife with a 5" blade, not more than 1" wide, edge to spine. If your patch knife is larger or smaller, test the pattern around your knife and see how well it fits. Make the pattern larger or smaller accordingly. Trace the pattern onto the leather with a soft pencil or awl.

Figure 160. Cut carefully with a sharp razor knife or shears.

Figure 161. A nice touch is to add a "pinked" or serrated edge folded over.

Antique patch knife with brass guard.
Author's collection.

162. Cut the decorative edging with a buttonhole chisel, or razor knife.

163. Dampen the sheath with water to soften it, and fold in half.

Figure 162. Cut the decorative edge with a buttonhole chisel or razor knife.

Figure 163. Dampen the sheath and fold in half, rough side to rough side.

Figure 164. Lay out stitch marks with an over-stitch wheel. Starting at the fold at the sheath tip, continue around the edge straight across the tabs, ending where the tab meets the decorative edge.

An antique J. Russell *Green River* knife with antler handle.
Author's collection.

164. Lay out the stitches with an over-stitch wheel.

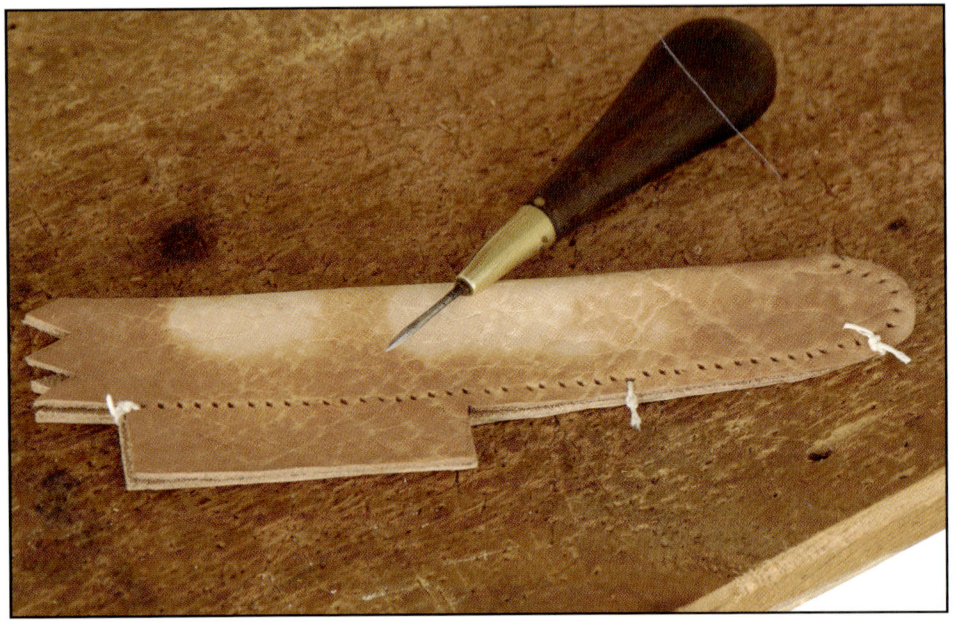

165. Punch the stitch holes using an awl. Tie three keeper stitches.

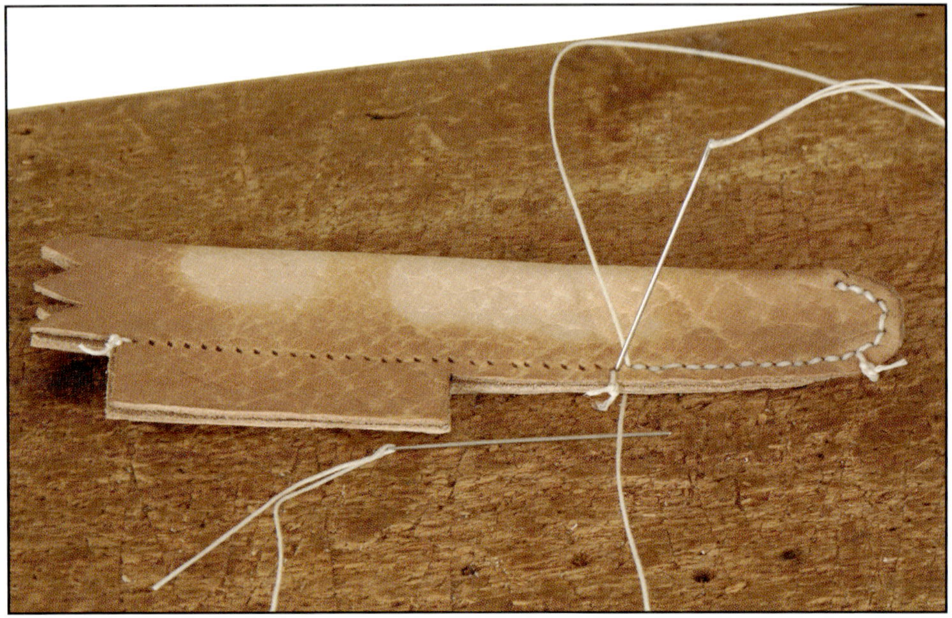

166. Saddle stitch the sheath.

167. Fold down the decorative top edge. Lay out stitches on the tab.

Figure 165. Punch three or more stitch holes with an awl. Tie keeper stitches to hold the folded sheath together, with keeper stitches in place, punch the remaining stitch holes with an awl.

Figure 166. Thread two needles, and saddle stitch the sheath. After stitching, carefully trim any uneven edges with a razor knife. Burnish the edges with a wooden awl handle.

Figure 167. Dampen the sheath, fold down the decorative top edge. Using an over-stitch wheel, lay out stitch marks along the edge of the extended tab, where the sheath will be stitched to the pouch. Insert your knife into the sheath and allow it to dry. Wrap a thin layer of plastic film, or put a thin layer of grease around the blade to keep it from rusting as the sheath dries.

168. Prepare the dye.

169. Dye the sheath with a sponge.

Figure 168. As the sheath dries, it will molded itself around the knife. Once completely dry, dye it to match the pouch and strap.

Figure 169-170. Dye the sheath using the same dye and techniques used on the pouch. Remember to dye the underside of the decorative edge. Dye an inch or more inside the sheath as well.

An antique patch knife, with square headed rivets.
Author's collection.

170. The dyed sheath.

171. Darken the sheath with shoe polish.

172. Remove the excess shoe black with a dry brush.

Figure 171-172. Apply brown shoe polish with a dauber. Buff it off with a dry boot brush. When complete, the sheath will match the strap, and be custom fitted to your patch knife.

Figure 173. The completed sheath, ready to be attached to the pouch strap.

173. The finished knife sheath.

Old patch knives with broken handles can easily be repaired with a rawhide wrap.

Author's collection

174. Position the sheath on the shoulder strap.

175. Refresh the stitch holes with an over-stitch wheel.

Figure 174. Try your pouch on as you intend to wear it. Locate the sheath on the outboard side of the adjustment strap, on the center of your chest, placing the bottom of the tabs roughly 10" above the top of the pouch flap. Mark that spot with a soft pencil. The sheath tabs will sandwich the strap as it is stitched in place.

Figure 175. Place the sheath tabs on your pencil mark, and carefully refresh the stitch marks with an over-stitch wheel.

Figure 176. Punch the stitch holes with an awl, piercing both sheath tabs and the sandwiched strap. Make your awl holes perpendicular to the strap. Do not wander on the backside.

176. Punch the stitch holes with an awl.

177. Saddle stitch the sheath to the strap.

178. Back stitch to finish, and cut the thread flush with a knife.

179. The sheath attached to the shoulder strap.

Figure 177. Using two needles and thread, saddle stitch the sheath tabs to the pouch strap. Finish with at least two back stitches.

Figure 178. Trim the thread flush with a razor knife.

Figure 179. Apply a little brown shoe polish to the white stitches and brush the entire area to blend the finish. The completed patch knife sheath is shown attached to the shoulder strap.

An antique patch knife marked "Ontario Pacific".
Pasche collection.

Attaching a powder horn:

180. Temporarily tie on the front strap, and punch with an awl.

181. Temporarily tie on the back strap, and punch with an awl.

Figure 180. Placement of a powder horn is personal. The powder horns on many original pouches are located at the middle of the flap. You may find this uncomfortable. It can get in the way of opening the flap. The powder horn can hang above or below the flap. Remember, the straps must be long enough to allow filling a powder measure.

Orient your horn as it will hang on your pouch, tip forward. Take one horn keeper adjustment strap, place it around the neck of the powder horn, smooth side out, with at least 1" of strap overlap. Temporarily tie the strap in place and punch an awl hole through both layers of the strap, close to the horn.

Figure 181. Place the second adjustment strap through the staple, at the base of the horn, fold smooth side out, with 1" of overlap. Temporarily tie the strap, punch an awl hole through both layers of the strap close to the horn.

Figure 182. Remove the straps from the horn. Starting at the awl hole, using an over-stitch wheel, mark four stitch holes along each edge of the keeper straps.

182. Remove the straps, mark four stitch holes with an over-stitch wheel.

183. Punch the stitch holes with an awl.

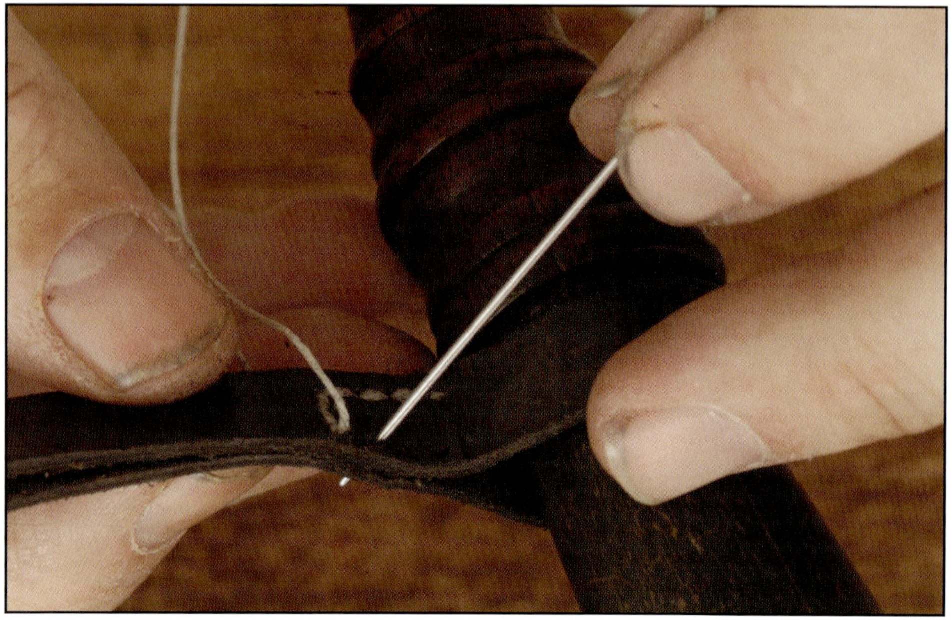

184. Attach the straps to the horn, saddle stitch the ends together.

185. Cut any excess leather with a razor knife or shears.

Figure 183. Punch the stitch holes with an awl through both layers of the strap, including holes to connect the two runs. Punch the awl holes perpendicular to the strap.

Figure 184. Replace the powder horn keeper adjustment straps on the horn. Saddle stitch the strap layers. Start near the horn and work around the stitch holes ending at the horn. Finish with at least two back stitches, and trim the thread flush with a sharp knife.

Figure 185. Trim any excess strap from the back side, above the stitches, with a knife, shears, or buttonhole chisel. Do not cut the finished horn strap. Use brown boot polish to darken the thread to match the pouch.

Antique scissor bullet mold.
Author's collection.

The finished pouch.

Installing a gusseted seam:

186. Pouch body with the welts and gusset.

187. Mark the stitches on the body with an over-stitch wheel.

188. Line up the body, gusset, short welt, and add a keeper stitch.

Figure 186. You will find that many original hunting pouches are made with an integral gusset. If you wish to add this feature to your pouch, here's how.

Cut the gusset, and short welt. The gusset for the North Hampton pouch can be found in the pattern section of this book. The gusset will be reverse stitched to the bottom of the body panels in between the long and short welts, one to face each panel. In this case, the gusset length, tip to tip, is about one half the length of the entire back panel measured around the edge, or 11-1/4" by 1-1/2" at the widest, gradually tapering to a point at either end. The short welt is 3/8" wide, and 1/2" to 1" longer than the edge of the gusset.

Figure 187. Dampen the rough side of the back panel and lay out stitches around the edge with an over-stitch wheel.

Figure 188. Find the exact center of the bottom edge of the back body panel. You may fold it to find the center. Mark the center by punching a hole with an awl along the row of stitches. Find the center along the edge of the gusset and punch an awl hole. Repeat for the short welt. Note that this welt needs to be longer than the gusset, 11-1/2" or 12" in length.

189. Add several temporary keeper stitches.

190. Punch the stitch holes with an awl.

Figure 189. Place the rear body panel smooth side up. Lay the short welt on the body panel smooth side down. Lay the gusset smooth side down on top of the short welt. Tie a keeper stitch through the awl hole, centering the three pieces. It is important that everything be centered now, or the finished pouch will be off center. Be careful to keep the body panel, gusset and sandwiched welt lined up. Punch and tie at least three more keeper stitches equidistant on each side of the centered keeper stitch.

Figure 190. Punch the stitch holes with an awl, through the panel, short welt and gusset. Start and stop about two full stitch marks from the gusset tips. Keep the pieces aligned along the edge, and be sure the holes are through the center of the welt, not drifting out of alignment.

Figure 191. Saddle stitch the gusset, short welt, and back body panel together. Trim the excess short welt to match the tip of the gusset.

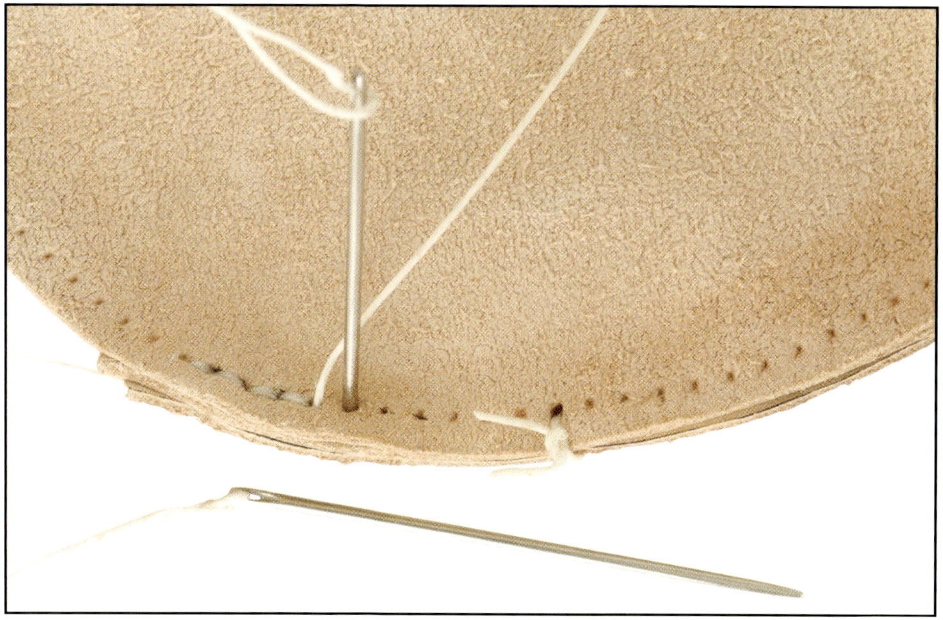

191. Saddle stitch the gusset, and short welt, to the back body panel.

192. Tie the front panel and long welt to the gusset with keeper stitches.

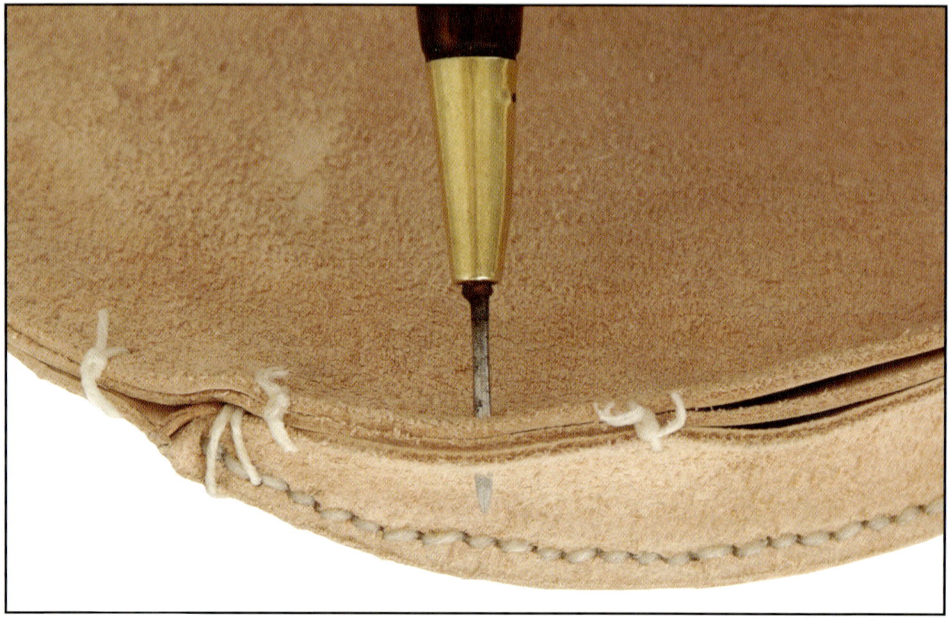

193. Punch the stitch holes with an awl.

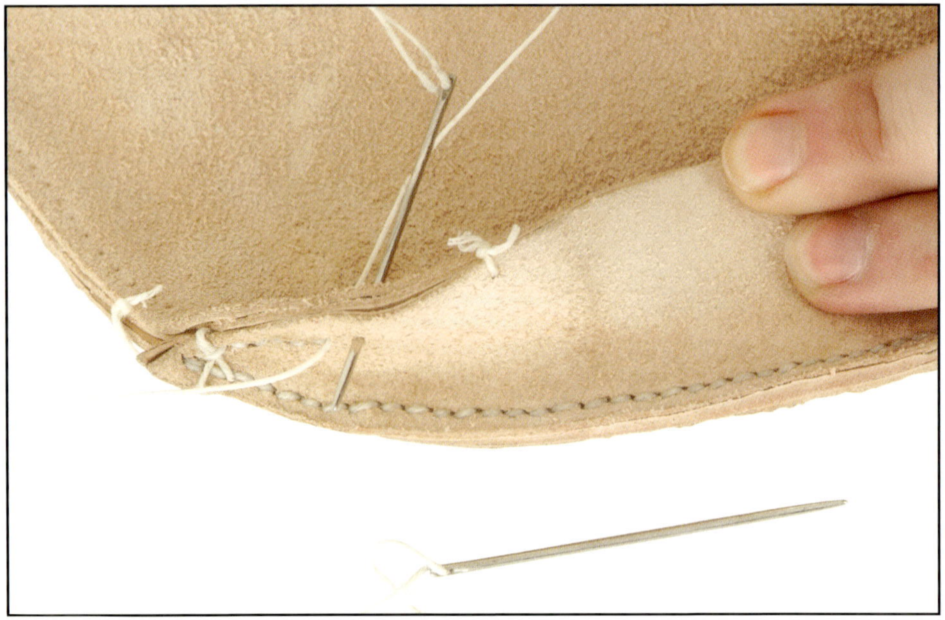

194. Saddle stitch the gusset and welt to the second body panel.

Figure 192. Dampen the edge of the front panel and lay out the stitch marks around the edge on the rough side with the over-stitch wheel. Find the center point along the bottom edge of the body and punch a hole with an awl. Mark the center of the long welt and gusset. The short welt was only as long as the gusset, the long welt will be long enough to line the entire edge of the body. Align the three awl holes on the front body panel, long welt, and gusset. Secure them with a keeper stitch. Add three more keeper stitches equidistant on either side of the center keeper stitch to the gusset. Add additional keeper stitches around the remaining body, securing the body pieces to the welt.

Figure 193. Punch the stitch holes along the edge of the front body panel where it adjoins the gusset only. Punch through both the front body panel, long welt, and gusset.

Figure 194. Saddle stitch the panel, welt, and gusset. Start and stop two full stitches from the tips of the gusset.

195. Saddle stitch the body panels.

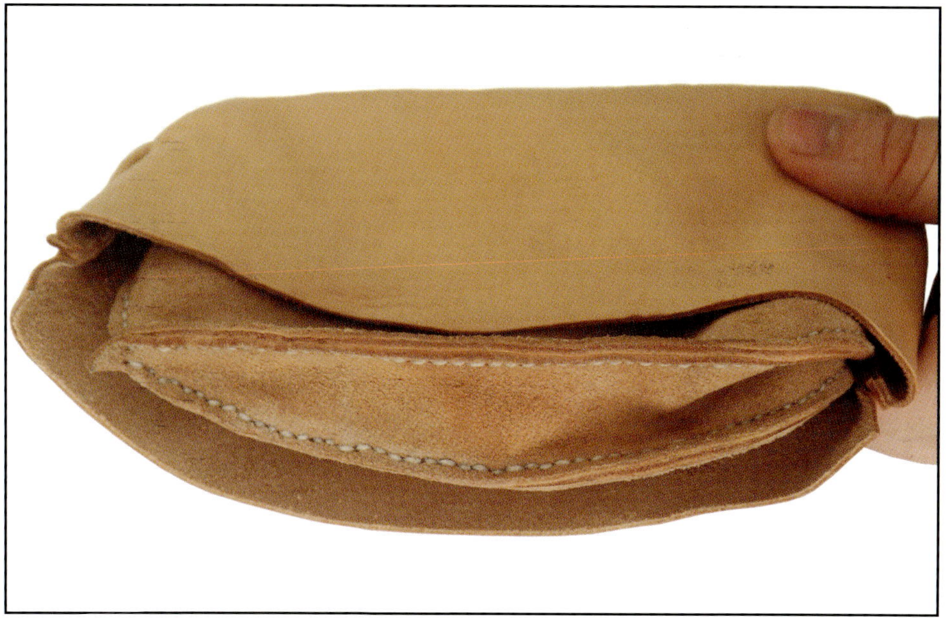

196. Soak the pouch in water to soften, and turn it smooth side out.

Figure 195. Punch the remaining awl holes along the edges of the body panels, keeping the long welt centered. Saddle stitch the body panels. Start at the top of the pouch stitching towards the gusset and ending at the gusset stitches. It's important to make the panel stitches tight to the gusset so there is no exposed gap. The point at the tip of the gusset will help insure that there is no gap, as long as you stitch tightly up to it. You may need to wet the leather to pull these stitches tight against the gusset. Back stitch this spot. It is a stress point. Trim any excess welt flush with the top edge of the back panel.

Figure 196. Soak the pouch in warm water and turn the pouch smooth side out. It should be easy to turn a pouch made of 3 oz. leather. If you used heavier leather and have trouble, pull the bag down over a broomstick handle to smooth out any hard to turn spots.

Figure 197. The welted gusset should appear as shown. Continue to finish the pouch as shown on pages 29-53 by adding the reinforcements, flap, and straps.

197. The finished gusset with welted seams.

Antiquing techniques:

You have already seen how to dye a leather bag. To add an attractive touch, some artists choose to antique their leatherwork, giving it an aged look as though it has been used for years. If used enough, your bag will attain this look on its own. There are many antiquing techniques, but here you will learn those that have worked best for me. You should feel free to experiment on your own. Some techniques should be applied before the item is dyed, and some after. Different types of leather may react differently to different dyes and antiquing techniques. I highly recommend experimenting on a scrap piece of leather before using on a finished project.

The antiquing techniques used here work best on un-oiled and un-waxed, dry vegetable tanned leather. You will need a spray type oven cleaner containing "lye", such as Heavy Duty Easy Off®, and your leather dye of choice. I prefer Feibings brand leather dye, in brown or black. You will also need a well-ventilated work space, preferably outdoors, when you spray on the oven cleaner or use other chemicals.

The double pouch shown here, like the previous pouch, is made of vegetable tanned leather. The pattern for this pouch can be found in the back of the book.

198. Wearing gloves and eye protection, spray the entire bag with Easy Off®.

199. Wearing gloves, wash the excess chemical off the pouch and dry.

200. When dry, apply the leather dye with a dauber.

Figure 198. When working with chemicals, be sure to wear gloves, eye protection, and work in a well ventilated area.

Heavily spray all of the smooth surfaces of the pouch, front and back, with the oven cleaner. The lye in the cleaner will remove the oils from the leather, causing premature aging and cracking. The leather will be permanently darkened where it contacts the cleaner.

Figure 199. Allow the cleaner to work a few minutes, then rinse the pouch in clean water. Wear rubber gloves to wash off the oven cleaner, as it will remove the oil from your skin causing severe dryness or chemical burning. Blot the leather surface with a paper towel and allow the pouch to dry completely, direct sunlight is best. The sun will accelerate the aging of the leather surface.

Figure 200. When totally dry, apply the leather dye as directed by the manufacturer. Directions should be located on the bottle. Alcohol based dyes will further artificially distress the surface of the leather.

201. Apply leather dye to the entire pouch, flap, and straps after aging.

202. On a double pouch, apply dye on the back sides.

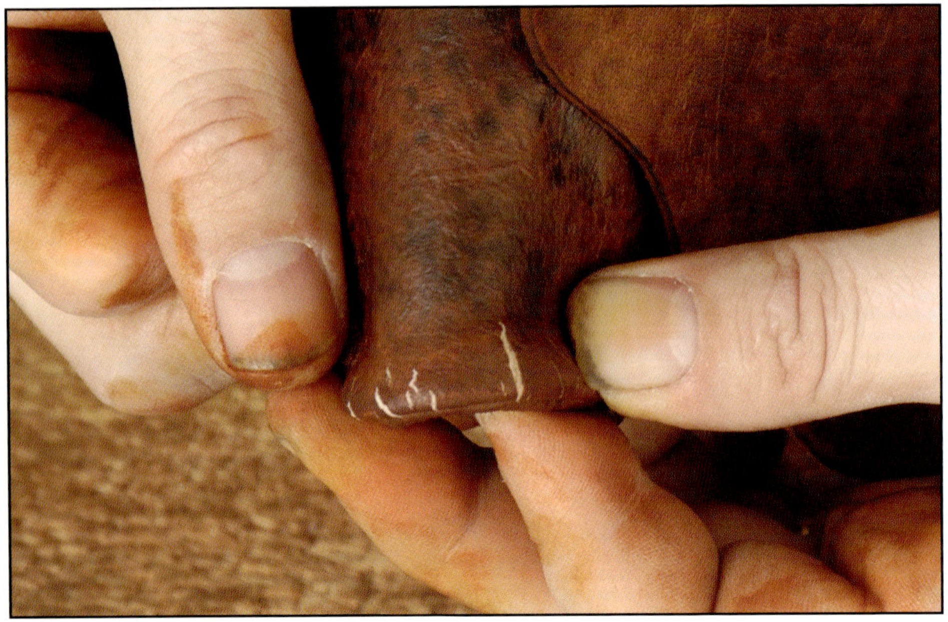

203. When dry, bend the leather to crack the edges.

Figure 201. Cover all of the smooth surfaces of the pouch and straps with leather dye. The dye dries quickly, cover any streak with repeated applications. Try to keep the color uniform and even as you apply it on the pouch.

Figure 202. This is a double pouch, dye the inner sides of the body as well. When dying is complete, allow the pouch to dry. The longer you wait, the better the next step will work.

Figure 203. The lye and dye have artificially dried the smooth surface of the leather to the point you can actually crack the surface. Make a sharp back bend on the edges of the seams, then roll the seam with a pushing motion. You will cause it to crack as shown. Bend and stress the strap, flap, seams, and body until you are satisfied with the results. It would usually take years of use to dry and crack the leather to this extent.

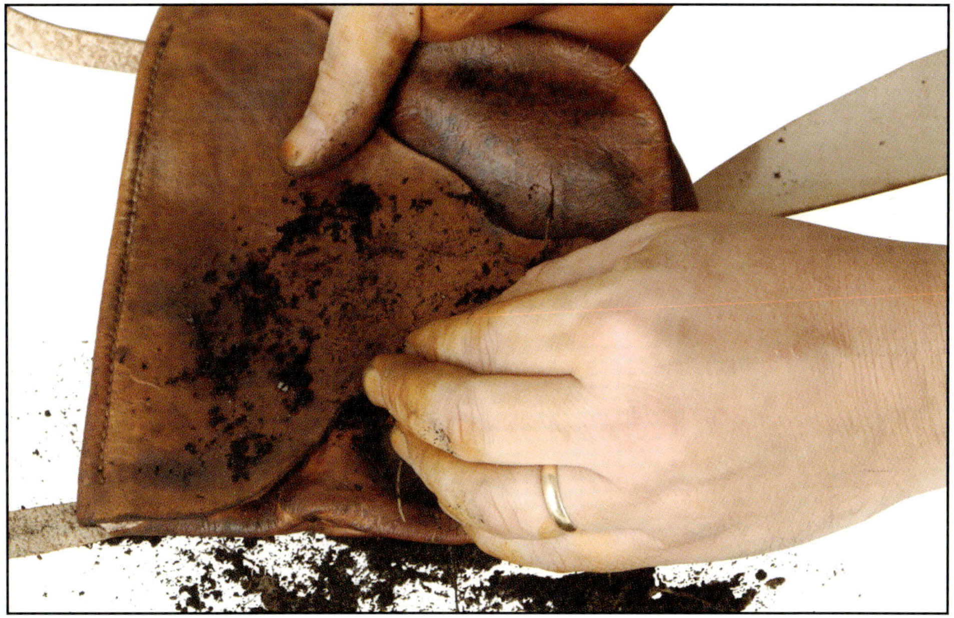

204. Apply a mixture of dirt and grease to age the pouch and straps.

Figure 204. An additional aging technique, giving a pouch a well used appearance, is to "dirty" the leather slightly. Some artists go to great lengths to simulate grime and dirt using paint, or other chemicals, I find it easiest and most effective to use the real thing. Simply take a small handful of pure, debris free garden dirt, and add a few teaspoons of vegetable oil, or melted lard. Mix it together into a black greasy paste. Smear the mixture of grease and dirt onto the surface, seams, and cracks of the pouch.

Rub dry dirt (without the grease) in the interior of the pouch to simulate use and age, paying special attention to any exposed stitches and seams. Apply brown shoe polish wax to seal everything, and buff with a boot brush.

The artificially aged hunting pouch.

Beaver tail pouch, c. 18th century.

Original antique hunting pouch
courtesy of A. Rehder

This 18th century pouch is made of black dyed leather and measures 7-1/2" long, by 7-3/4" wide, with a 11" long flap. The shoulder strap measures 2-1/8" wide, by 34" long, attached to the pouch with black leather lace. The pouch is stitched with a double welt that extends 1/4" to 3/8" beyond the edge of the pouch.

Small hunting pouch.

Original antique pouch
courtesy of A. Rehder

This simple black leather pouch measures 6-3/4" wide, by 5-1/2" tall. The button that once secured the 4-5/8" long flap is now missing. The 1-1/2" wide, 42" long shoulder strap is made of two pieces of leather spliced together. The shoulder strap is attached to the pouch at an angle, lessening the stress on the pouch when worn. The shoulder strap once had two powder horn straps, which are now missing. The iron buckle has three tongues. The welted seam is the edge of the second interior pocket.

Hunting pouch, c. 1770 - 1800.

The patch knife measures 6-7/8”, with a deer antler handle, and blunt point blade.

Original antique hunting pouch, circa. 1770-1800
courtesy of A. Rehder

About 8” by 8”, with a beavertail front flap, this pouch has one inside pocket, with no gusset.

The powder horn has a pyramid shaped neck, fitted with a leather stopper. The horn is 13-1/2”, with a 2-5/8” diameter pine plug.

The pouch has a woven or loomed strap, 39” long, spliced together at the top.

The back of the pouch has a leather sheath for the patch knife. Note how the woven strap is sewn at an angle on one side, and straight on the other.

1770 - 1800 era Colonial hunting pouch.

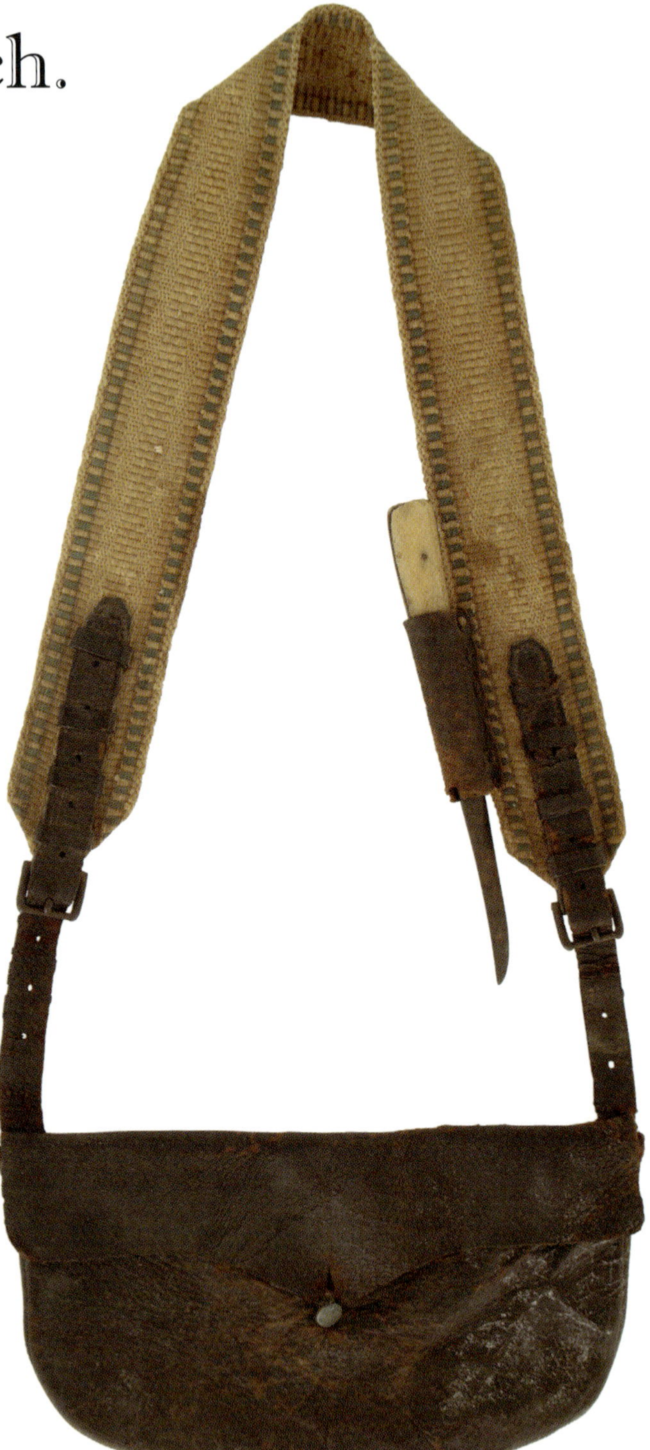

This 1770-1800 era hunting pouch is reminiscent of a military style cartridge pouch. The pouch is black dyed leather with a welted seam, measuring 10" wide, by 6" tall. The flap is secured by a small round brass button, with a loop on the back.

The loomed shoulder strap measures 2-1/2" wide. Two thinner leather adjustment straps, about 9" long with iron buckles, secure the pouch to the strap. A leather sheath for the patch knife is sewn to the shoulder strap. The patch knife has tapered tang with bone scales secured by two pins.

The two leather straps are sewn directly to the back of the pouch near each corner. The buckle strap has four leather strap keepers. The powder horn has a fluted chip carved domed plug and raised ring.

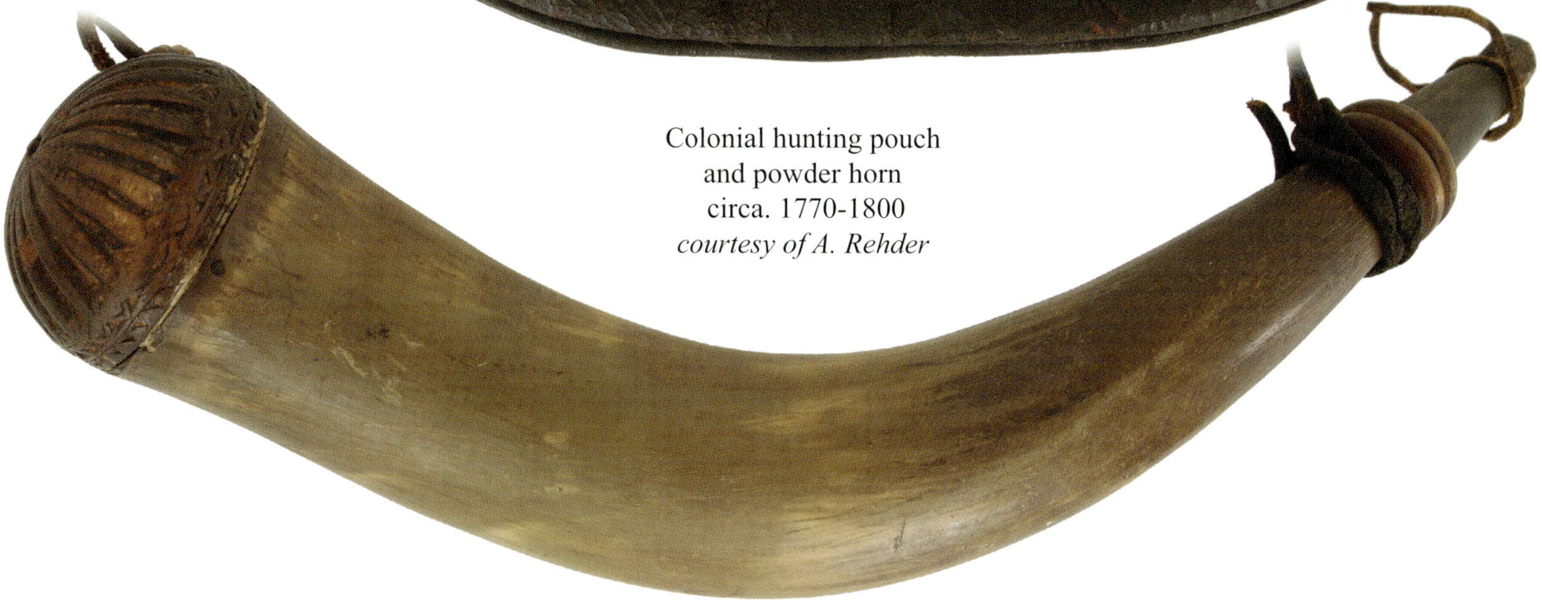

Colonial hunting pouch
and powder horn
circa. 1770-1800
courtesy of A. Rehder

Fleur-de-lis hunting pouch.

Original fleur-de-lis
antique pouch
circa. early 19th century
courtesy of A. Rehder

This early 19th century pouch measures 7-3/4" wide, by 6" long. The flap has a black dyed fleur-de-lis with stitched borders. The flap is secured with a leather tab (now missing), and a leather knot. The replacement shoulder strap is 44" long, and attaches to the pouch with two iron buttons. A 3" long bone powder measure is attached to the shoulder strap.

Cowhide pouch, c. early 1800's.

Original antique pouch
courtesy of A. Rehder

This pouch has a reddish brown cowhide front panel, and a black dyed leather back panel and flap. The pouch measures 10" wide, by 8-1/2" long, with a 2-3/4" wide bottom gusset. The 5" long flap closes with an iron button. The date 1800 or 1808 is stitched to the front of the flap, as well as a folk art bird.

Eastern Pennsylvania hunting pouch.

This hunting pouch from Eastern Pennsylvania, was found with a George Zollinger rifle. The pouch measures 8" by 8", made of black dyed deerskin. The flap and parts of the front body retain some original hair. The bottom of the bag has a gusset and welted seams, for added space and additional strength.

The 2-1/2" wide loomed shoulder strap measures 40" long, it is stitched to the top seam of the bag.

The powder horn measures 12-1/2", with a 2-1/2" diameter chip carved butt plug with iron staple. The neck is carved octagon-to-round with a raised ring. The horn hangs almost to the bottom edge of the flap.

Original Eastern Pennsylvania hunting pouch,
circa. late 18th century to early 19th century
courtesy of A. Rehder

Deerskin double pouch, c. early 1800's.

The front panel and flap of this pouch are deerskin, tanned with the hair on. The edges are bound with a 3/8" wide strip of leather, boastsing 11 stitches per inch. The pouch measures 9" wide, by 6-3/4" long, with a 7-3/4" long flap.

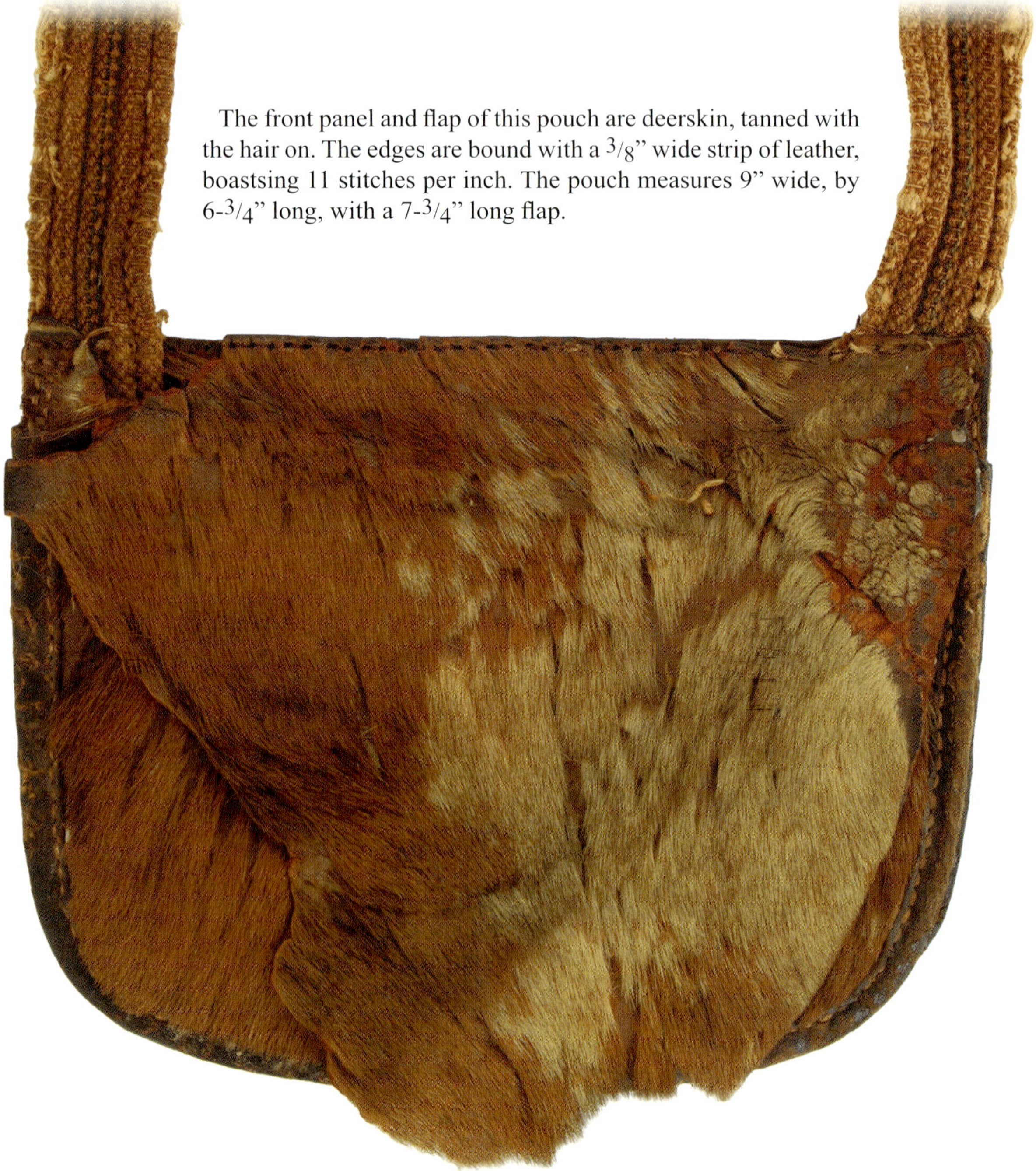

Original antique pouch
courtesy of A. Rehder

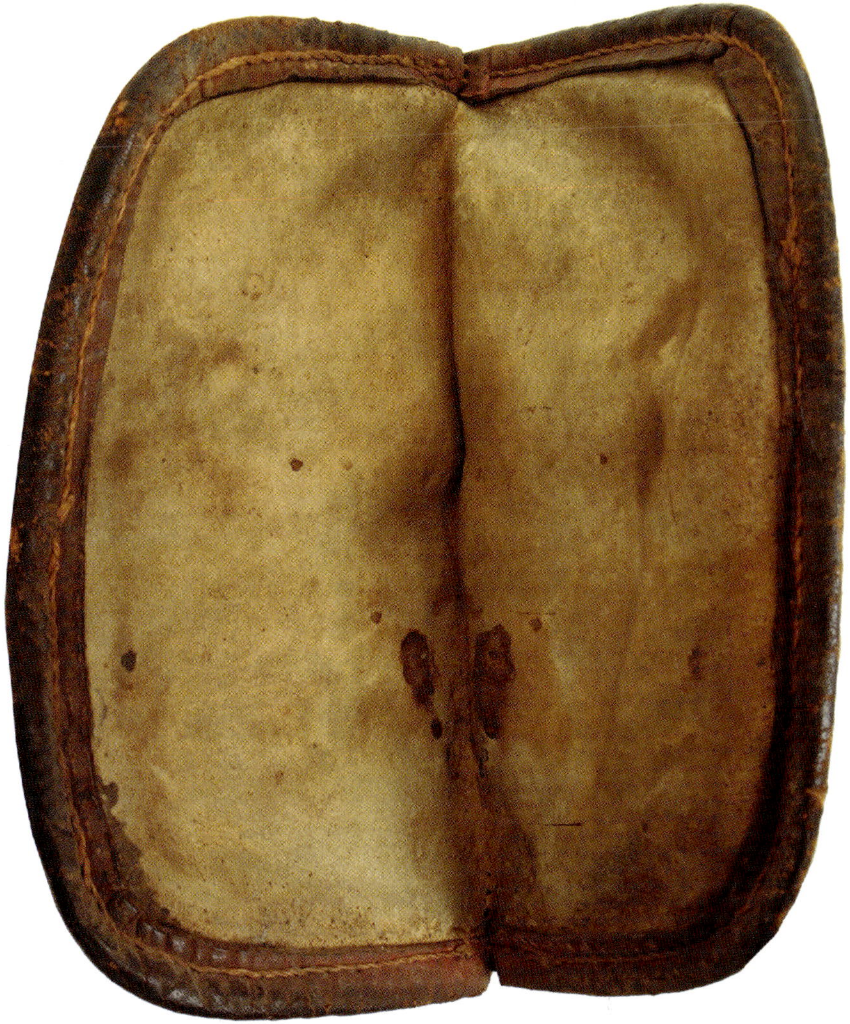

A view between the two pockets.

The woven shoulder strap is 1-1/8" wide, by 38" long. The ends of the shoulder strap are sandwiched between the front and back panels of the pouch.

Beaver tail hunting pouch.

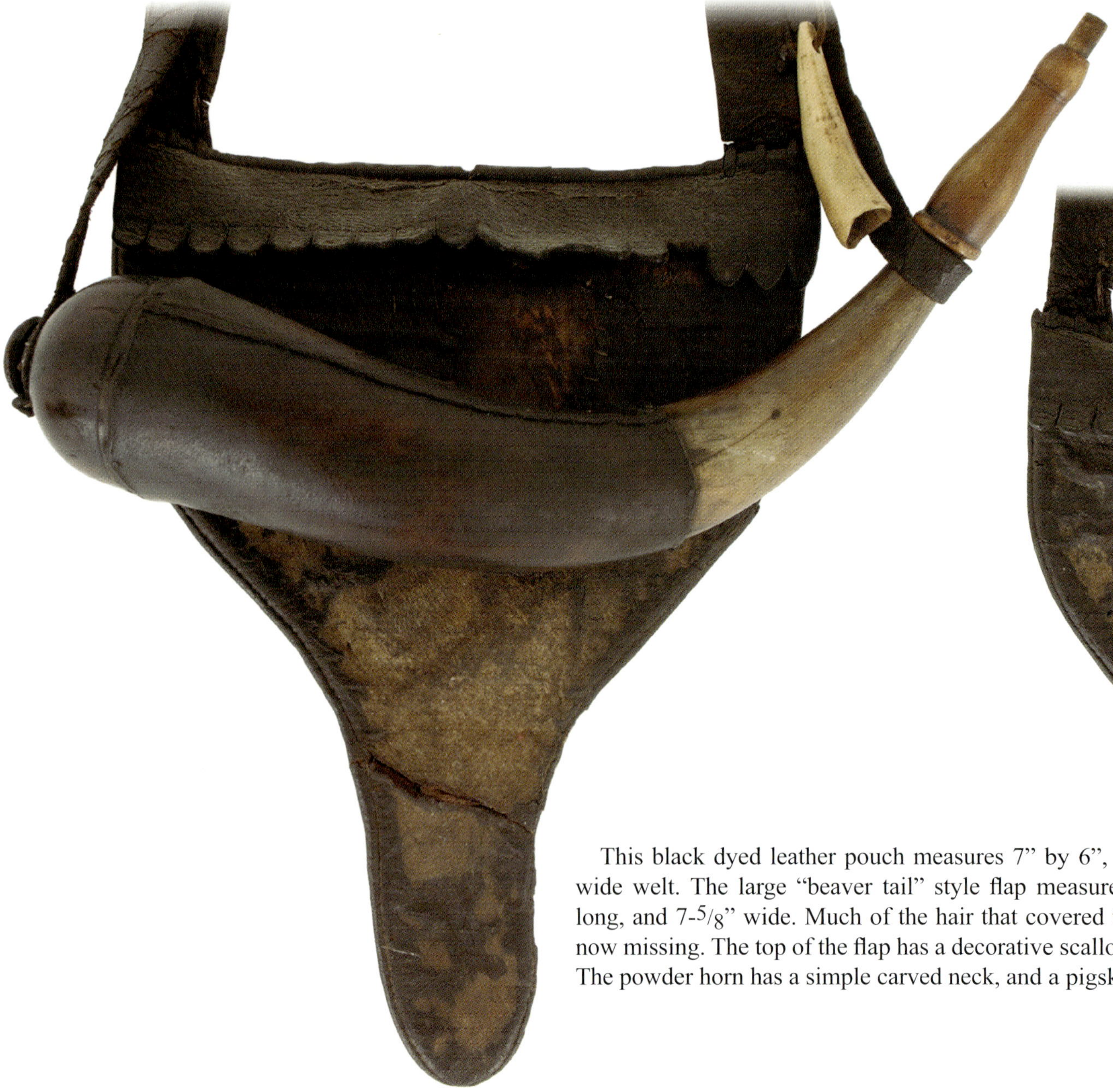

Original antique beaver tail hunting pouch
courtesy of A. Rehder

This black dyed leather pouch measures 7" by 6", with a 2" wide welt. The large "beaver tail" style flap measures 10-1/4" long, and 7-5/8" wide. Much of the hair that covered the flap is now missing. The top of the flap has a decorative scalloped edge. The powder horn has a simple carved neck, and a pigskin repair.

An antler powder measure is attached to the shoulder strap with a black leather cord. The shoulder strap is attached to the pouch with thick black leather lace. The shoulder strap is 40" long, and has six adjustment holes spaced 3" apart, adjusted with an iron buckle. The powder horn measures 11-$^{1}/_{2}$" long, with a 2-$^{1}/_{2}$" diameter butt plug.

Hunting pouch, c. 1820.

This simple hunting pouch came from Tennessee and dates around 1820. The pouch is made of buff colored leather, measuring 9-1/2” wide, by 8-1/2” long. The 17” long powder horn has a screw tip, and 2-3/8” diameter turned butt plug. The pouch and horn are for right hand use.

Original 1820 hunting pouch
courtesy of A. Rehder

The back panel of the pouch is made of black dyed leather, with rows of stitching. The shoulder strap is made of black dyed leather, measuring $^{7}/_{8}$" wide, by 41" long. The shoulder strap has several thinner straps for attaching the powder horn, antler powder measure, and other accouterments.

Left hand double pouch, c. 1820 - 1840.

This double pouch is made of black dyed leather, 9-1/2" wide by 8" deep. The flap has simple geometric tooling and a serrated top edge. The flap, and body have bound edges stitched with a heavy linen thread. The double pouch begins half way down the middle of the bag, shown in the side view.

The factory made powder horn measures 12-1/4", with a turned screw-tip. The 2-3/16" diameter butt plug is stamped "J B SMITH, Lancaster". Only a handful of other horns are known to bear this stamp.

Left hand double pouch
circa 1820-1840
courtesy of A. Rehder

The 1-$^{3}/_{8}$” wide shoulder strap is stitched inside the pouch at an angle, making it more comfortable to wear. The 41” long shoulder strap is adjusted with an iron buckle, located on the front strap.

Double hunting pouch.

This kidney shaped double pouch is made of black dyed leather, measuring 8-1/2” wide by 7-1/2” long. Only small patches of brown hair remain which once covered the front flap. The flap measure 5-3/8” long. The 13-1/2” long powder horn has a flat pine butt plug, and a simple raised head. The initials ‘NH’ are engraved on the front side. The pouch and horn are for a right handed wearer.

Original antique double hunting pouch
courtesy of A. Rehder

The flap is lined with a blue and white ticking fabric. The raw edge of the flap is covered with leather. The 40" long shoulder strap has three adjustment holes, adjusted with a brass buckle located on the rear strap. An antler powder measure is attached to the shoulder strap.

Deerskin pouch, c. 1830 - 1840.

Original antique pouch
courtesy of A. Rehder

This curved pouch has a two piece deerskin front panel, and a black dyed leather back panel and flap. The pouch measures 6-$^{1}/_{2}$” wide, by 6-$^{3}/_{4}$” tall, with a 3” long flap. The welted seam has decorative piping. Two powder horn straps are attached to the $^{7}/_{8}$” wide shoulder strap.

Hunting pouch with horn, c. 1830 - 1850.

Original antique hunting pouch
courtesy of A. Rehder

This hunting pouch is made of black dyed leather, and measures 7-1/2” wide, by 6-7/8” tall. The 3-5/8” long flap is decorated with 1/2” long fringe. The replacement shoulder strap is 9/16” wide, and 34” long. The powder horn is original to the pouch, and measures 11” long, with a 2-1/2” diameter butt plug.

Hunting pouch, c. 1820 - 1840.

This circa. 1820 to 1840 leather hunting pouch measures 9” by 9”, with a 7-1/4” long bound edge flap. The sides of the pouch are gusseted to allow for more interior space. About three quarters of the strap has been replaced with 1/2” wide harness leather.

The die formed brass powder measure is lead soldered along the seam, embossed with a chevron pattern. A wad of tow (cotton waste) is inserted to reduce the powder charge. The measure hangs from a thin leather cord tied to the buckle. It measures .415” diameter, about 2” in length, plus the lanyard loop. Shown actual size.

The powder horn measures 13” long, with a 2-3/4” diameter butt plug, affixed with wooden pegs.

1820 to 1840 era hunting pouch
courtesy of A. Rehder

Beneath the flap, the front of the pouch has a small leather pocket with flap, possibly for holding percussion caps. The interior has one unlined pocket. The split seam gusset is lined with coarse woven canvas.

The $^1/_2$" wide leather shoulder strap is stitched to the upper back panel of the pouch, and adjusts with an iron buckle. The shoulder strap is a late 19th century replacement.

Initialed hunting pouch.

'CR' initialed
original antique pouch
courtesy of A. Rehder

This hunting pouch measures 7-$^{1}/_{2}$" long, by 7" wide, with a 4" long flap. The initials 'CR' are cut out on the flap with leather backing. The edges are decorated with fringe. The shoulder strap has two matching buckles. A powder horn with turned tip, a brass charger, and a simple wire vent pick are attached to the shoulder strap.

Fringed double pouch.

This fancy double pouch is made of black leather, measuring 7-1/2" wide, by 6-5/8" long. Two fringed layers of leather are stitched between the pouches. The top of the 4" long flap has a double serrated edge. The flap has bound edges. A sleeve is sewn to the pouch for holding a 4" long brass powder measure. The sliding iron adjuster has hand cut numbers.

Original antique pouch
courtesy of A. Rehder

The shoulder strap is a late 19th, or early 20th century replacement. The pouch has quality stitching, with 11 stitches per inch. The top corners of the pouch have leather reinforcements.

Tawed Leather Pouch, c. 1845 - 1850.

This tawed leather pouch measures 9" long, by 7" wide, with a 5-1/2" long flap. The black dyed shoulder strap is a late 19th century replacement. The powder horn hangs above the flap, so as to be out of the way when opening the pouch. Tawing is a mineral tanning process involving Alum (aluminum potassium sulfate).

Author's collection

Hunting pouch, 1886.

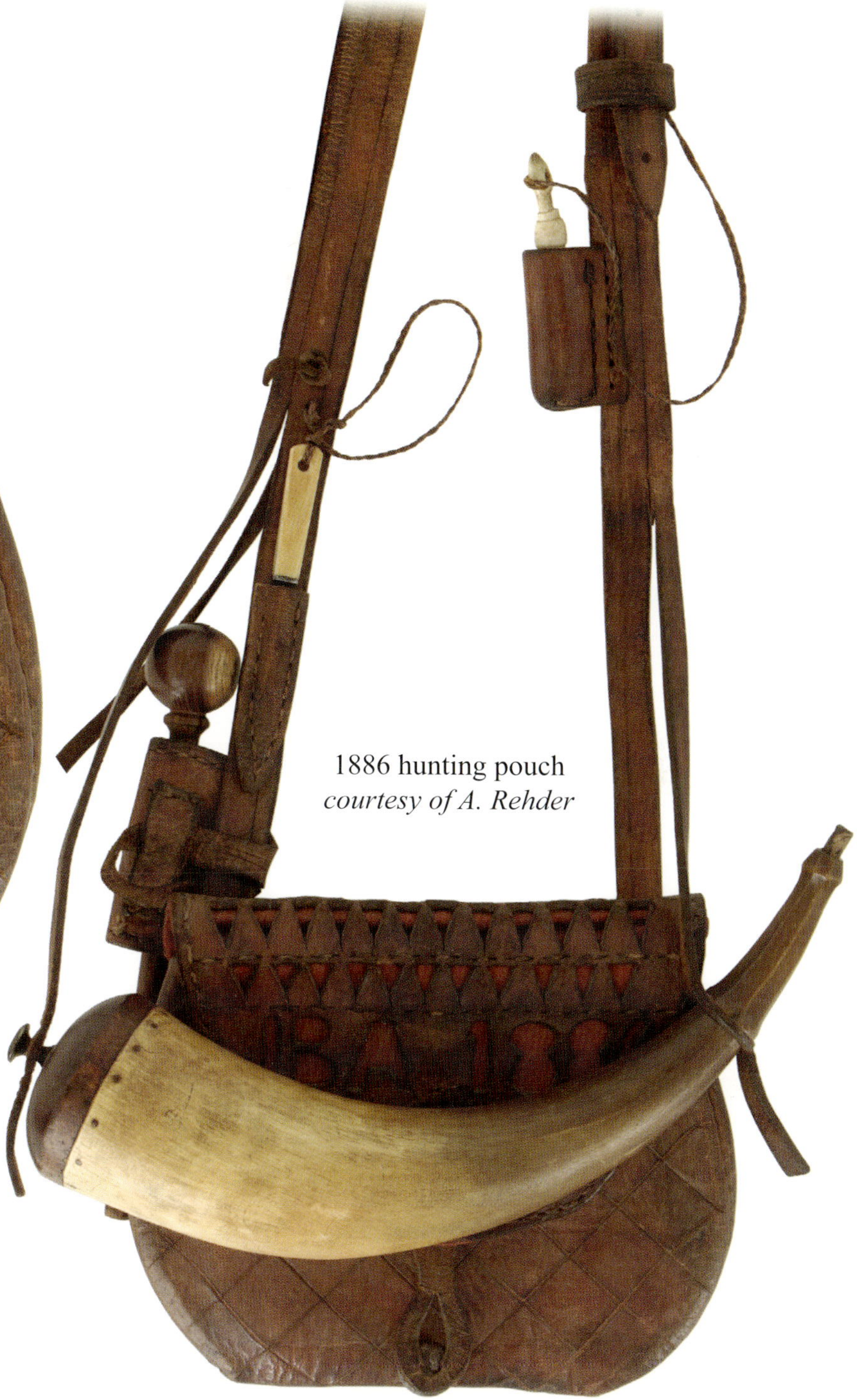

1886 hunting pouch
courtesy of A. Rehder

This pouch, dated 1886, measures 8" by 6-1/2" long. The body of the bag has a cut diamond pattern. The pierced flap has geometric and floral designs, 'J.B.A. 1886' initials and date, over a red wool trade cloth fabric. The flap is secured with a leather loop and U.S. military eagle brass button. Attached to the 44" long shoulder strap is a leather pocket for holding a bone powder measure, short starter or axe, and a bone handled vent pick. The 12-1/2" long powder horn has an octagonal carved neck and a domed butt plug.

The open view of the pouch shows the decorative score marks adorning the front panel, and the orange wool lined flap. A small leather pocket is sewn to the back of the pouch. The slot which now holds a short starter may have originally been used to hold an axe.

Still attached to the pouch is a carved antler powder measure, and an awl style vent pick with a bone handle.

North Hampton pouch.

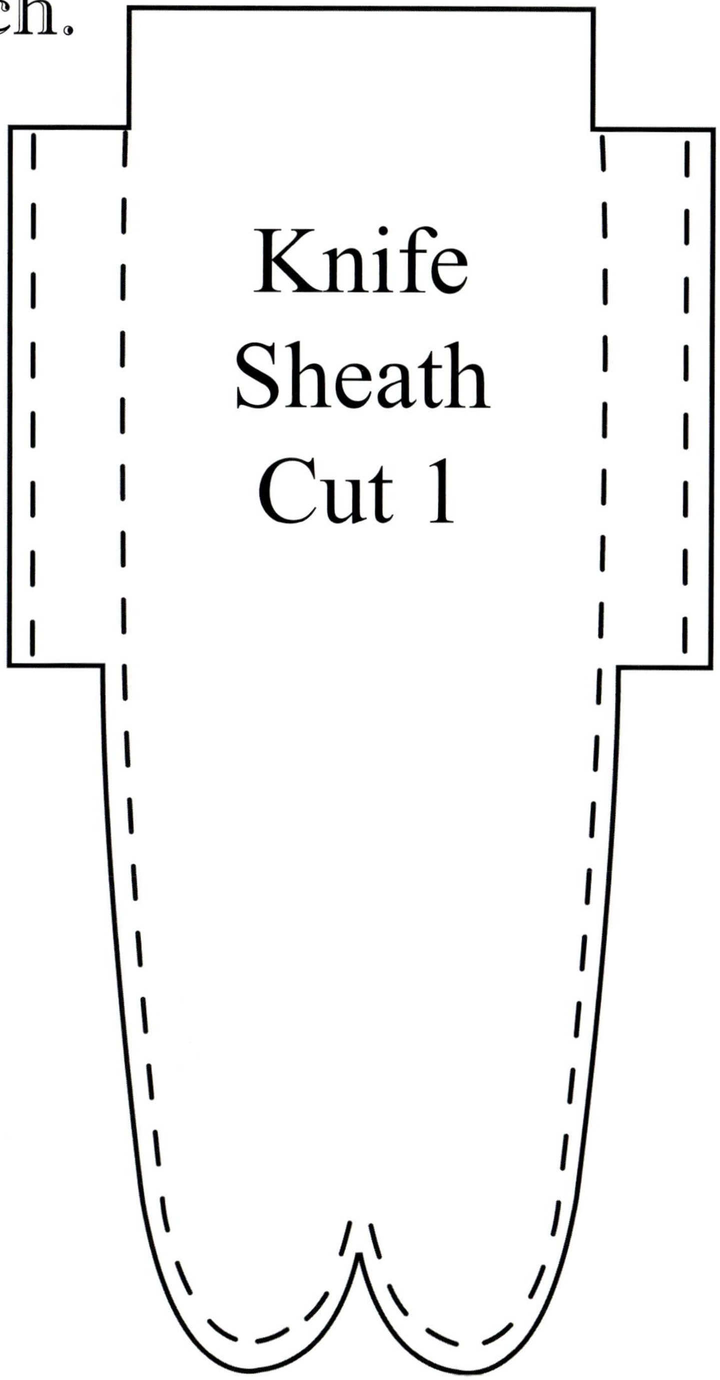

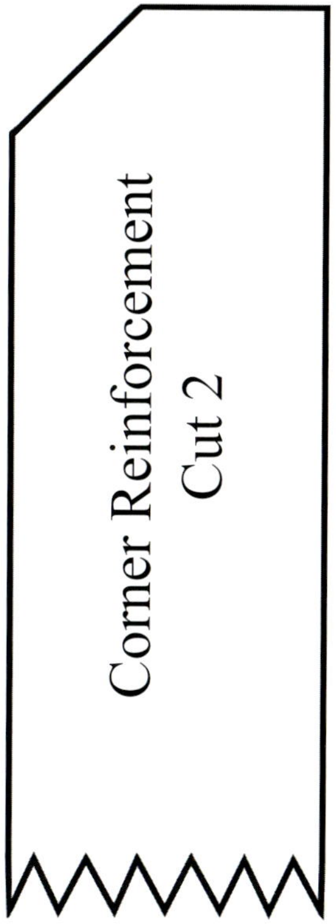

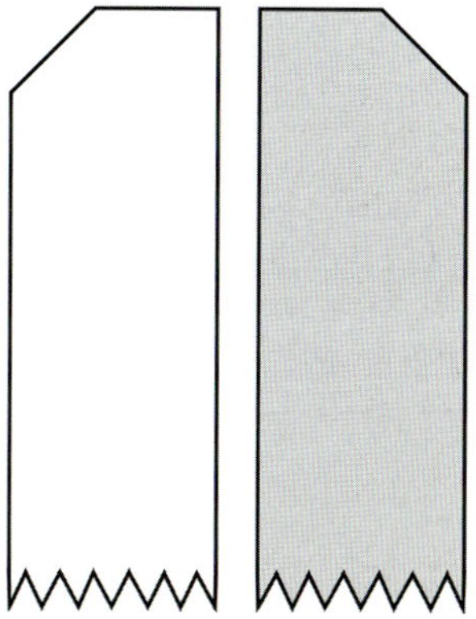

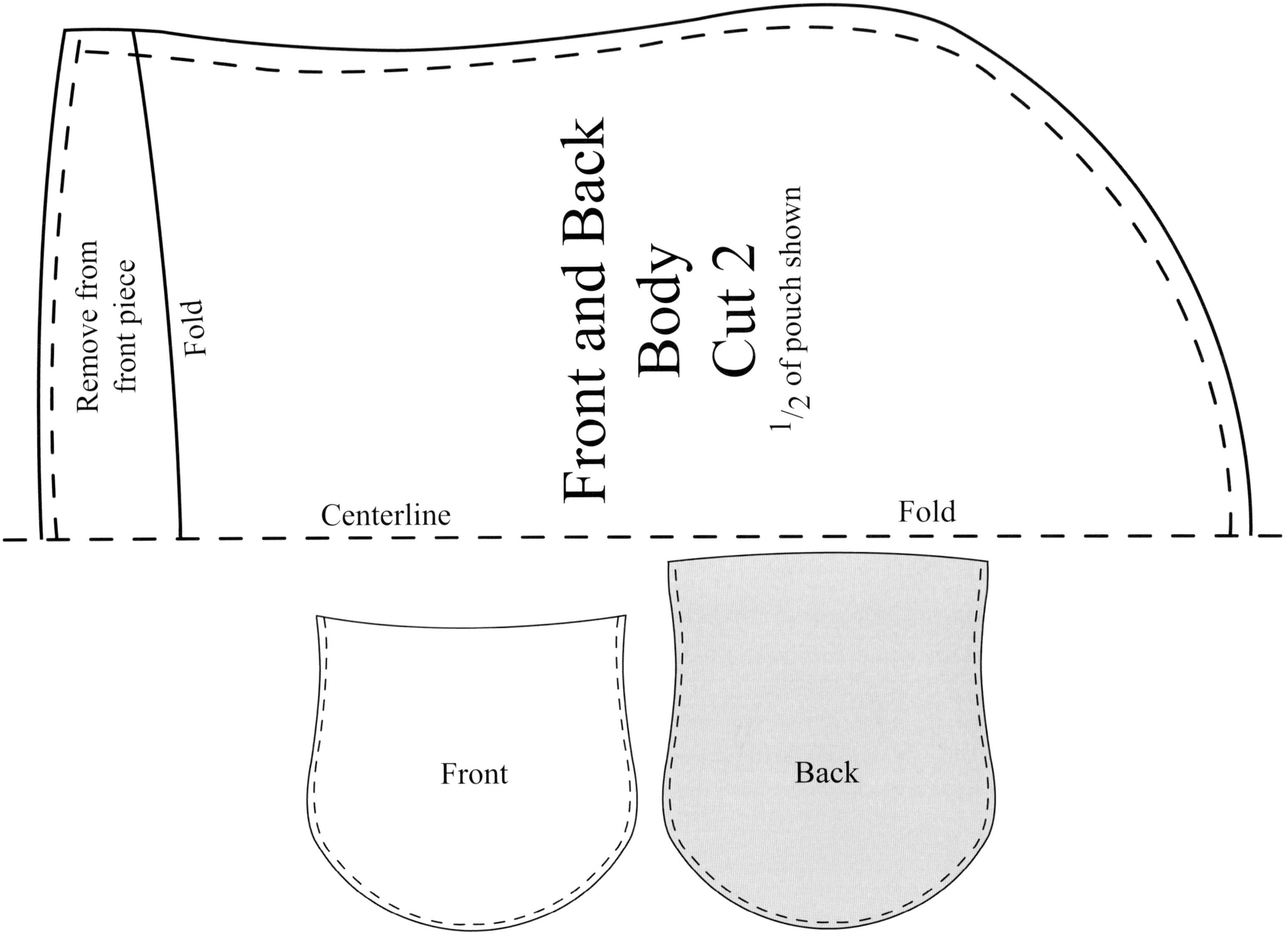
Remove from
front piece
Fold
Front and Back
Body
Cut 2
1/2 of pouch shown
Centerline
Fold
Front
Back

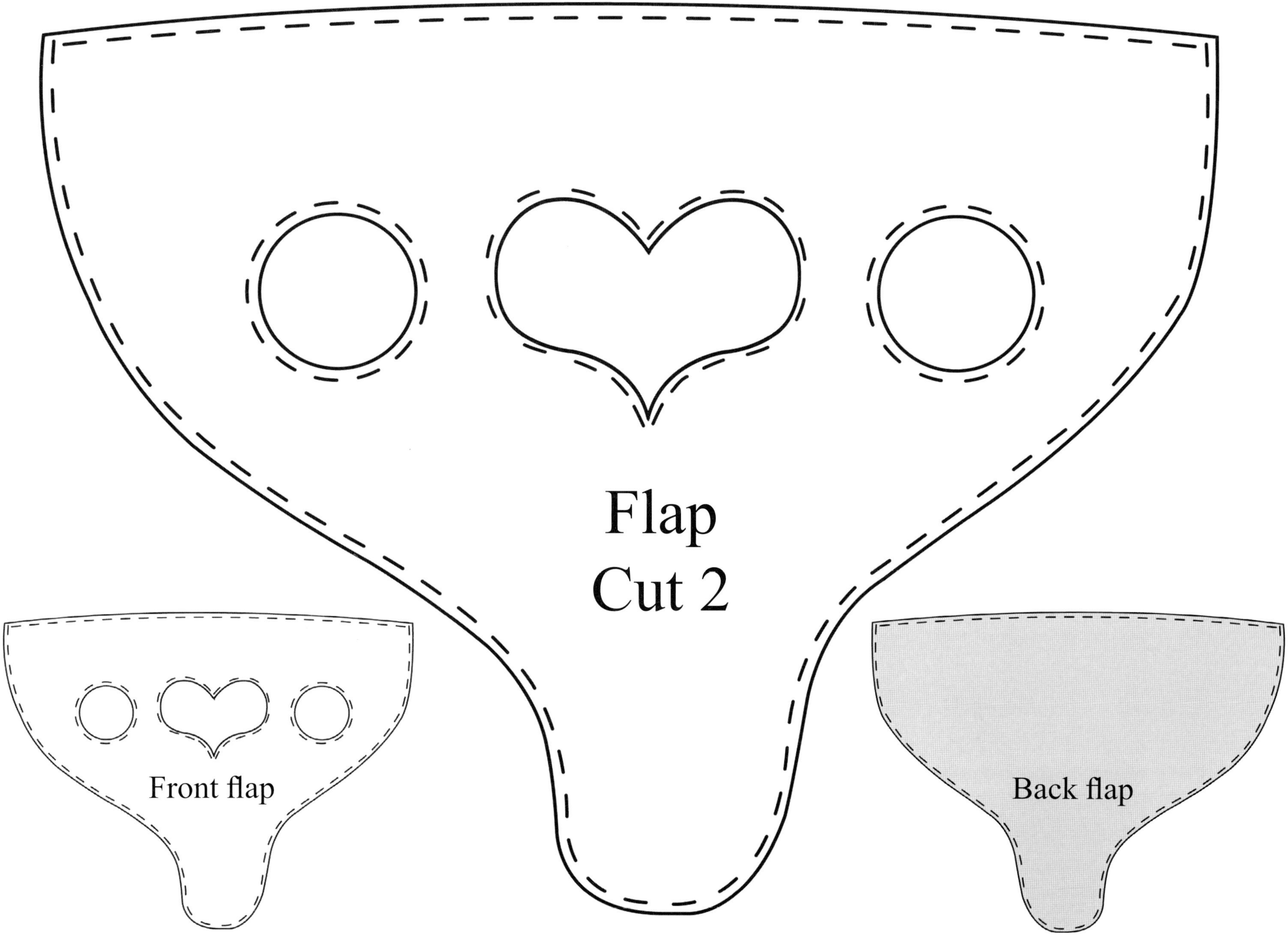
Flap
Cut 2
Front flap
Back flap

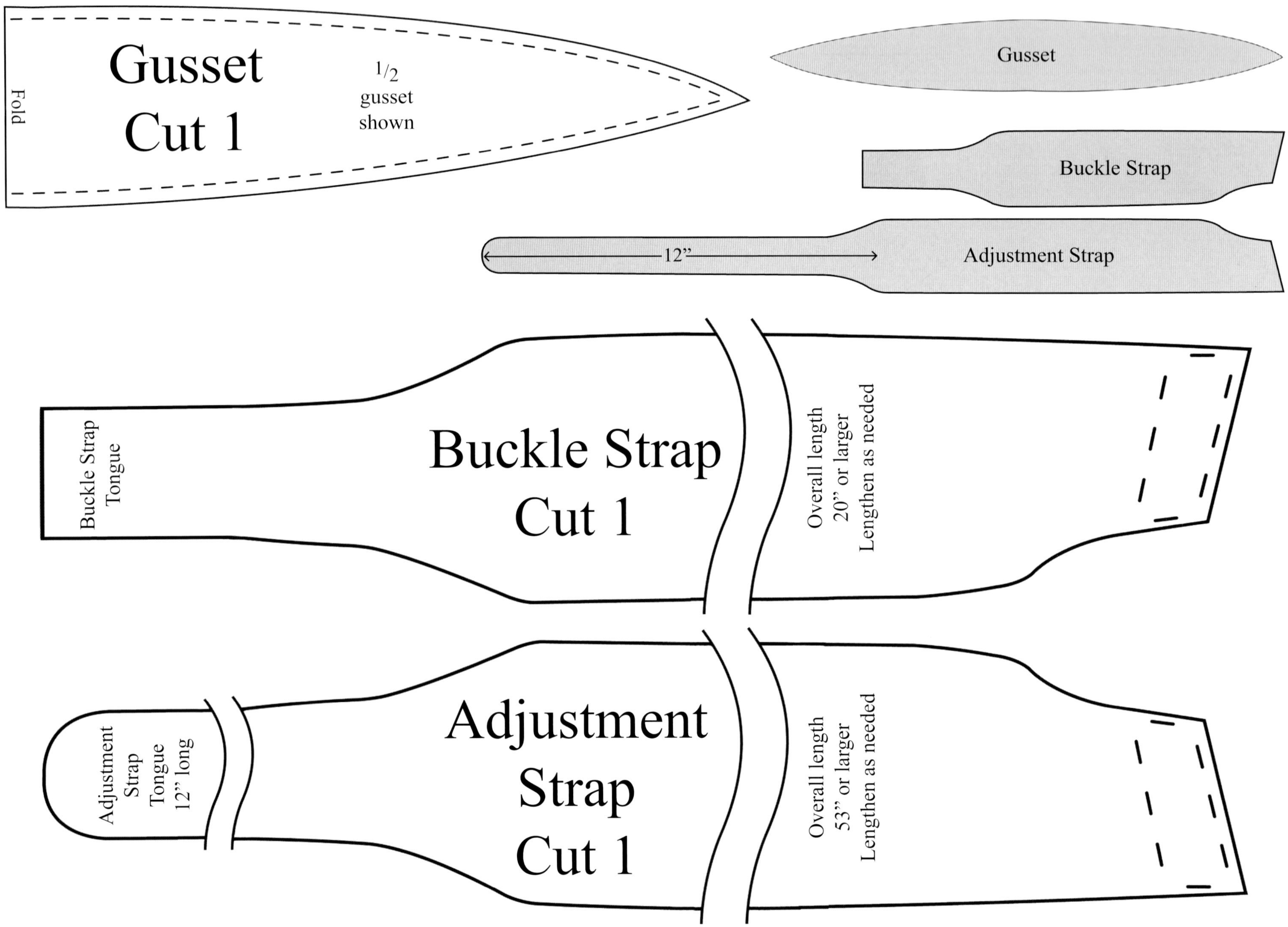
Gusset
Cut 1
1/2
gusset
shown
Fold
Gusset
Buckle Strap
12"
Adjustment Strap
Buckle Strap Tongue
Buckle Strap
Cut 1
Overall length
20" or larger
Lengthen as needed
Adjustment Strap Tongue 12" long
Adjustment
Strap
Cut 1
Overall length
53" or larger
Lengthen as needed

Double shooting pouch.

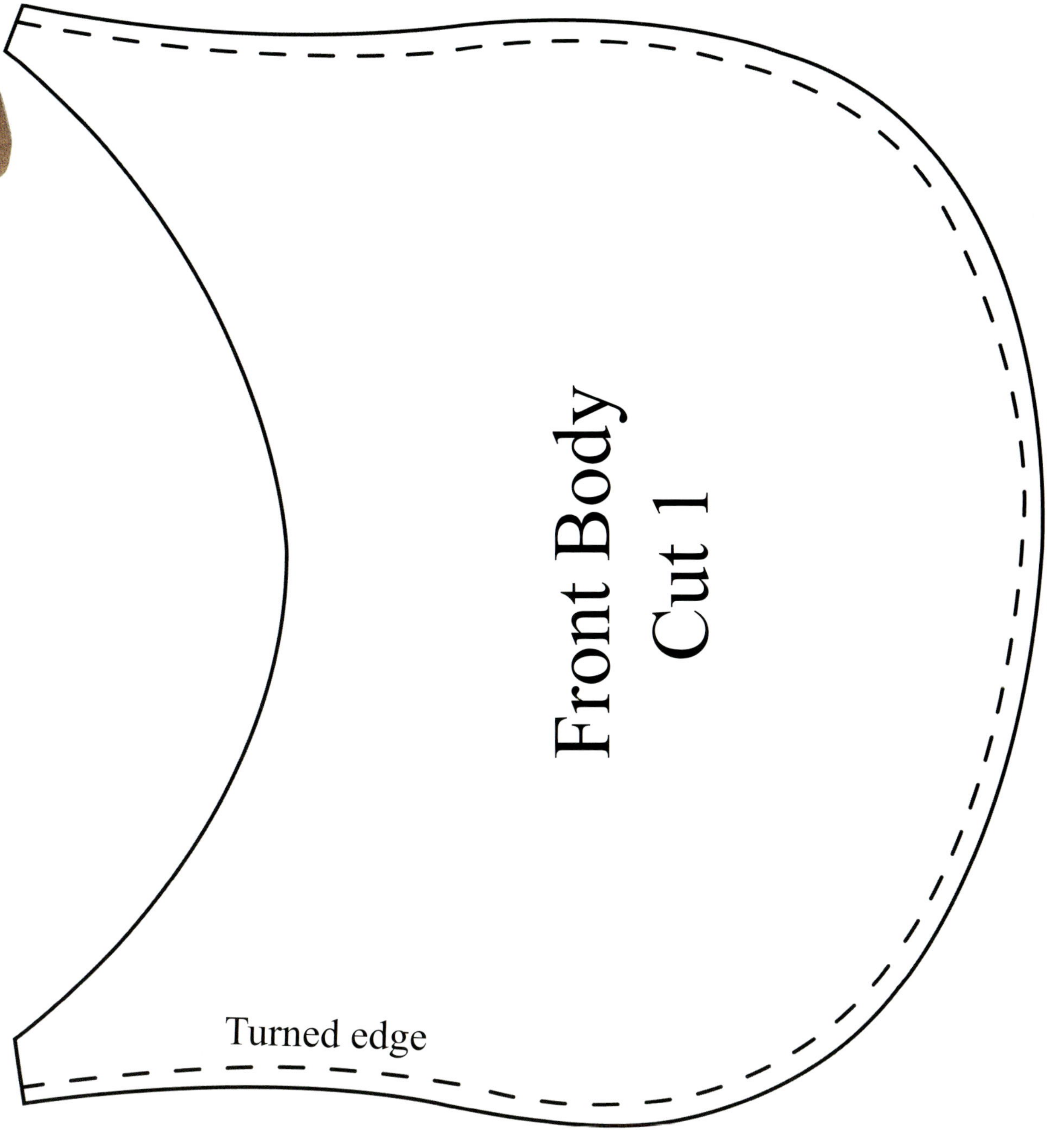

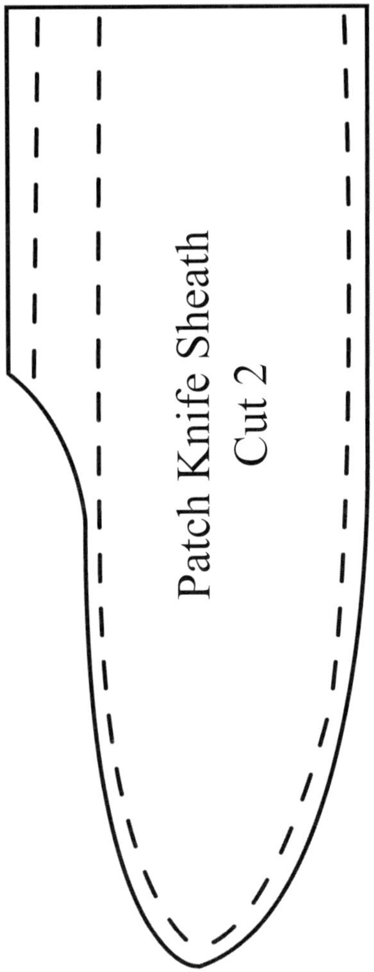
Patch Knife Sheath
Cut 2

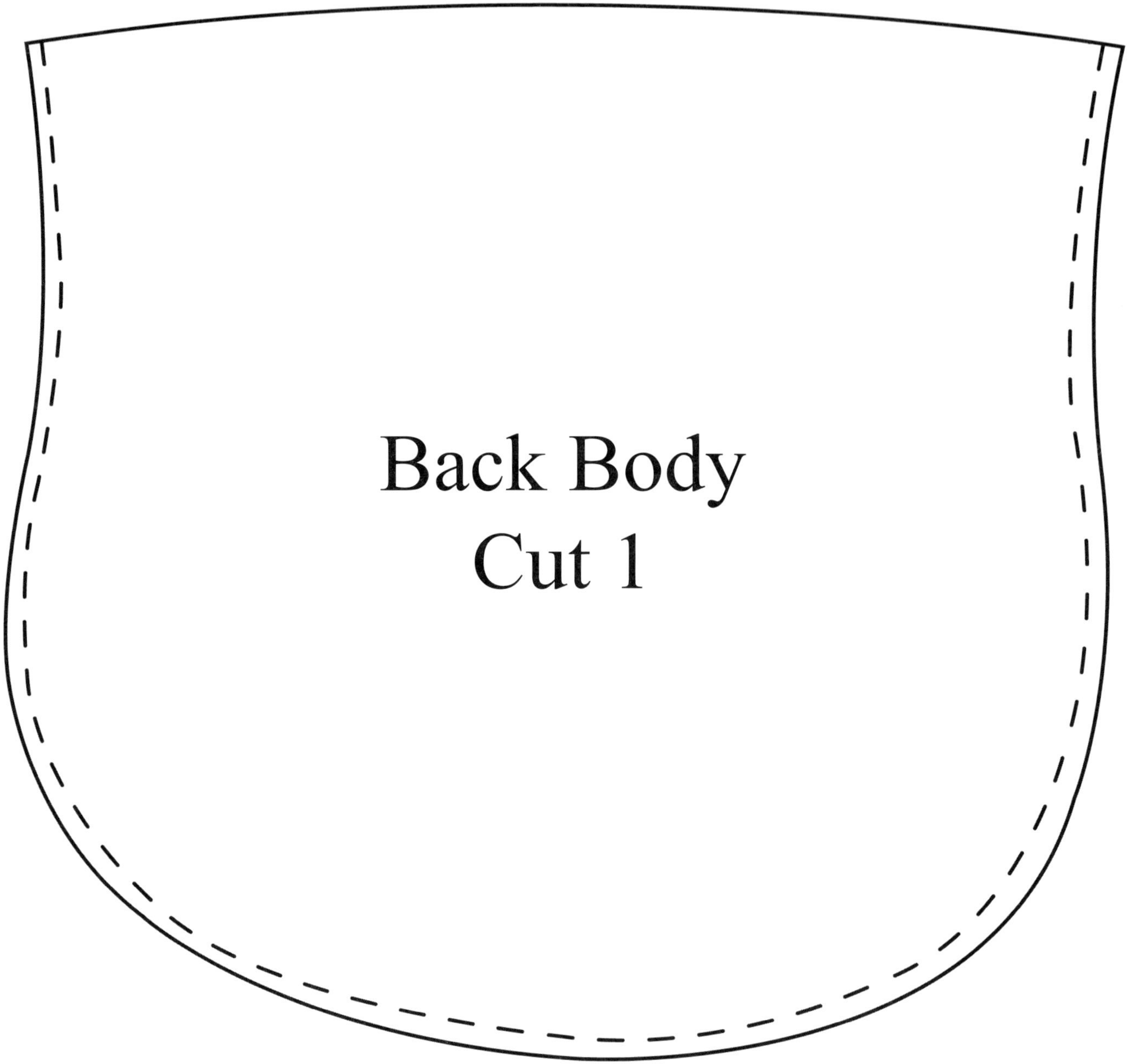
Back Body
Cut 1

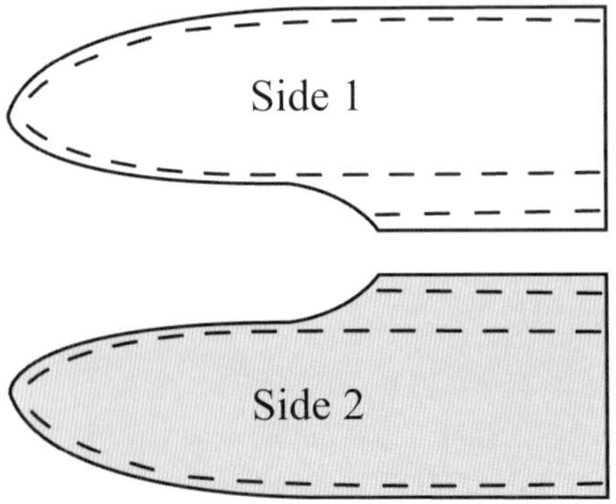
Side 1
Side 2

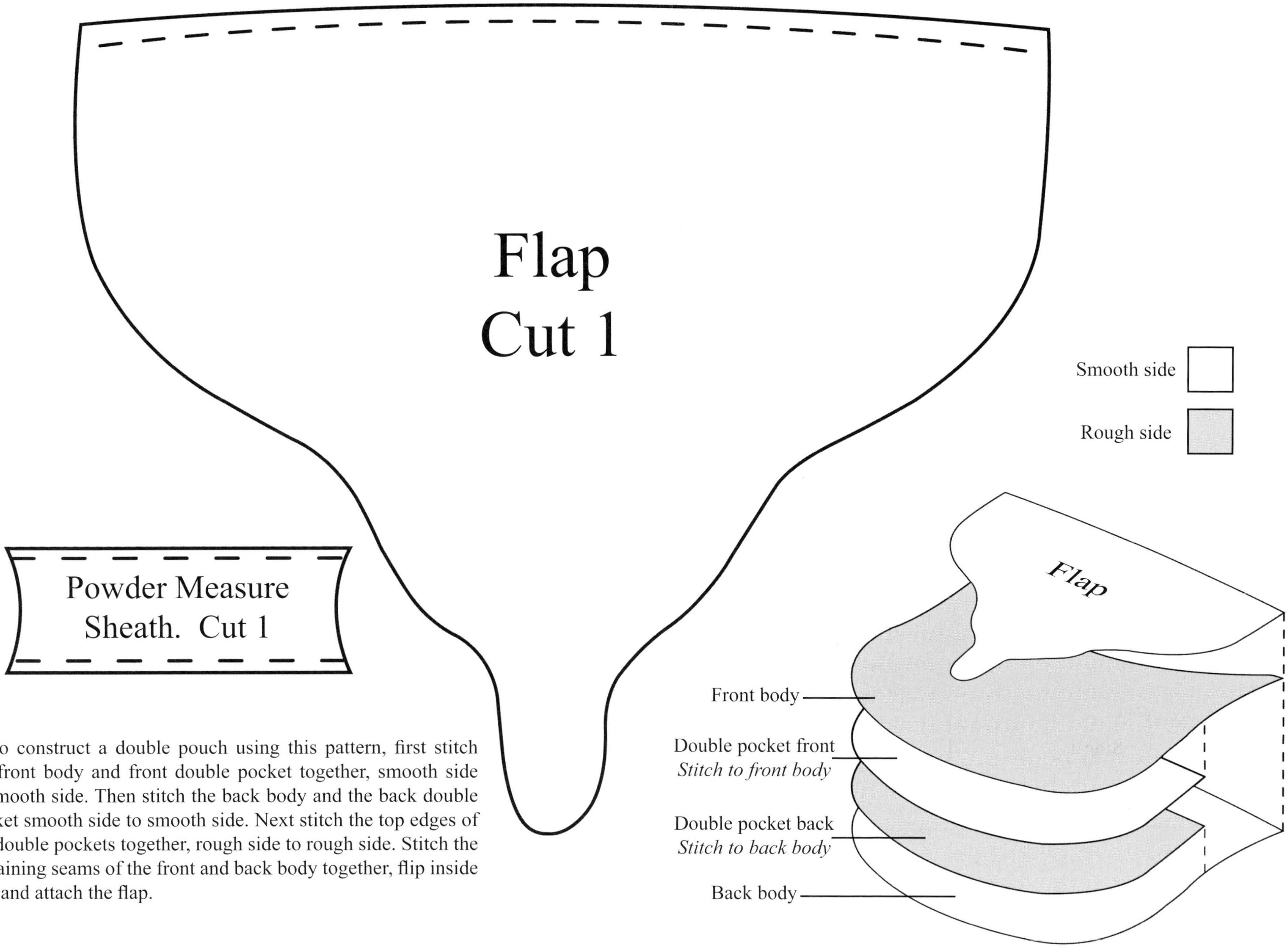

To construct a double pouch using this pattern, first stitch the front body and front double pocket together, smooth side to smooth side. Then stitch the back body and the back double pocket smooth side to smooth side. Next stitch the top edges of the double pockets together, rough side to rough side. Stitch the remaining seams of the front and back body together, flip inside out, and attach the flap.

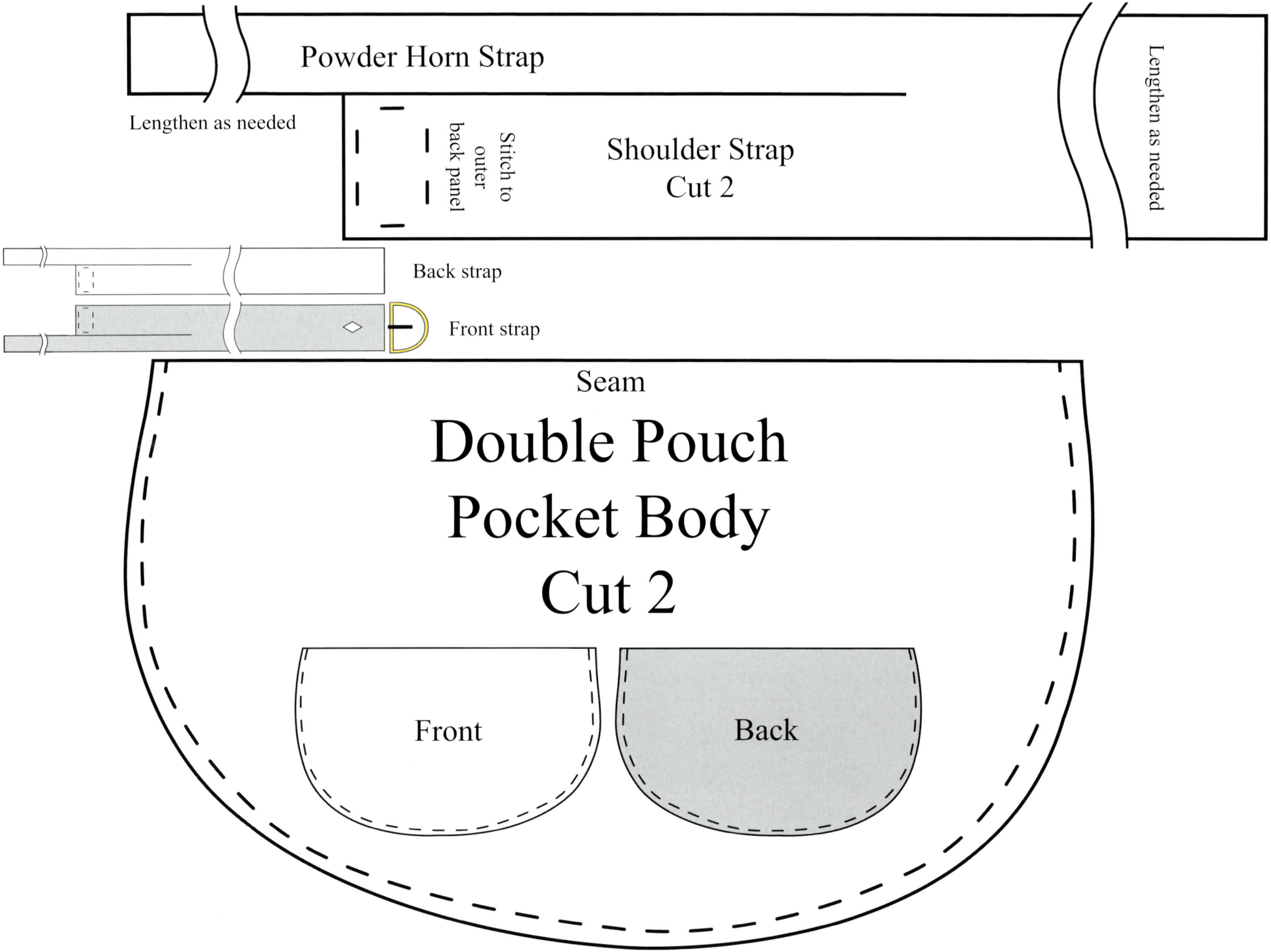
Powder Horn Strap
Lengthen as needed
Stitch to outer back panel
Shoulder Strap
Cut 2
Lengthen as needed
Back strap
Front strap
Seam
Double Pouch
Pocket Body
Cut 2
Front
Back

Tennessee Rifle Bag.

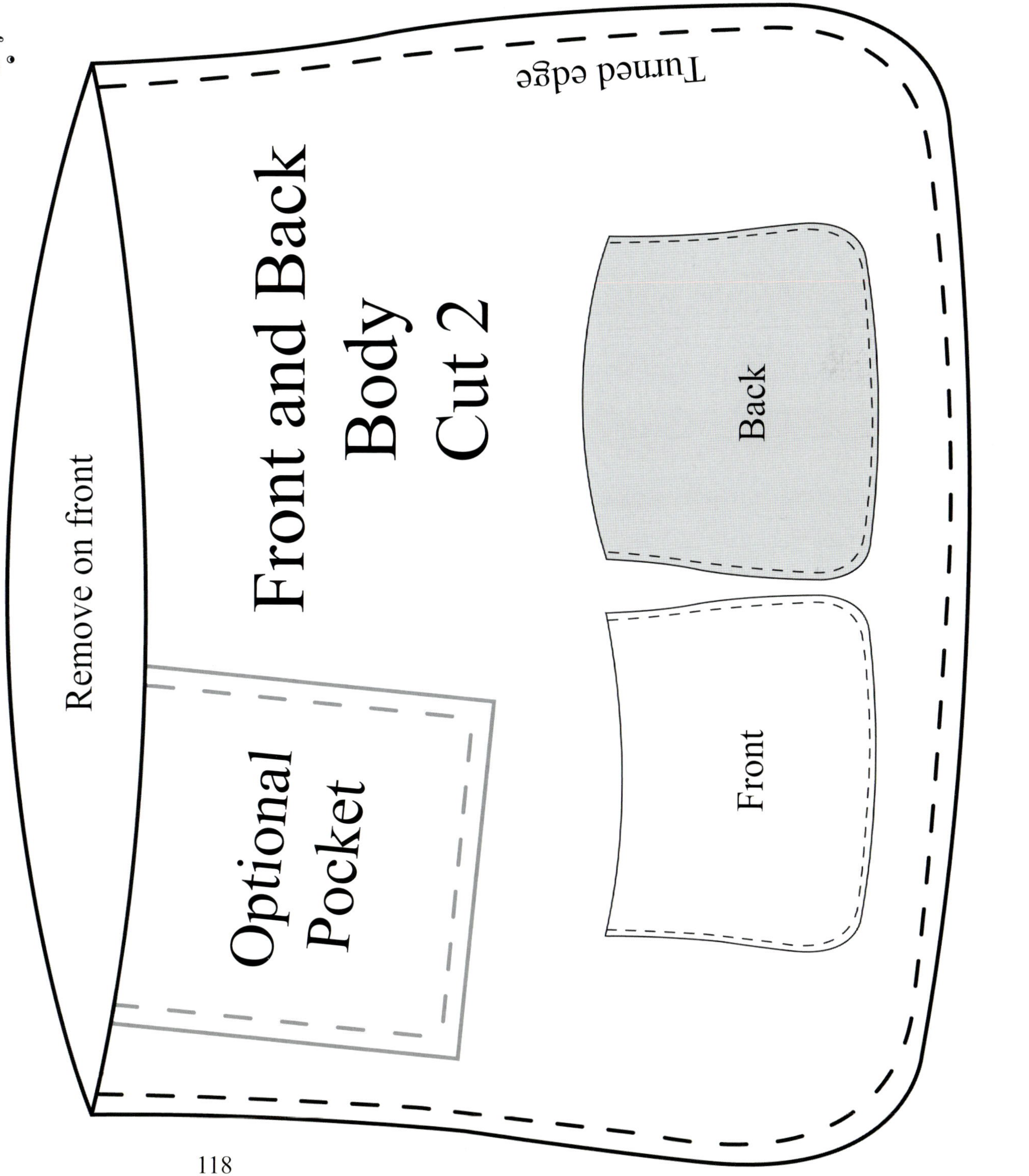

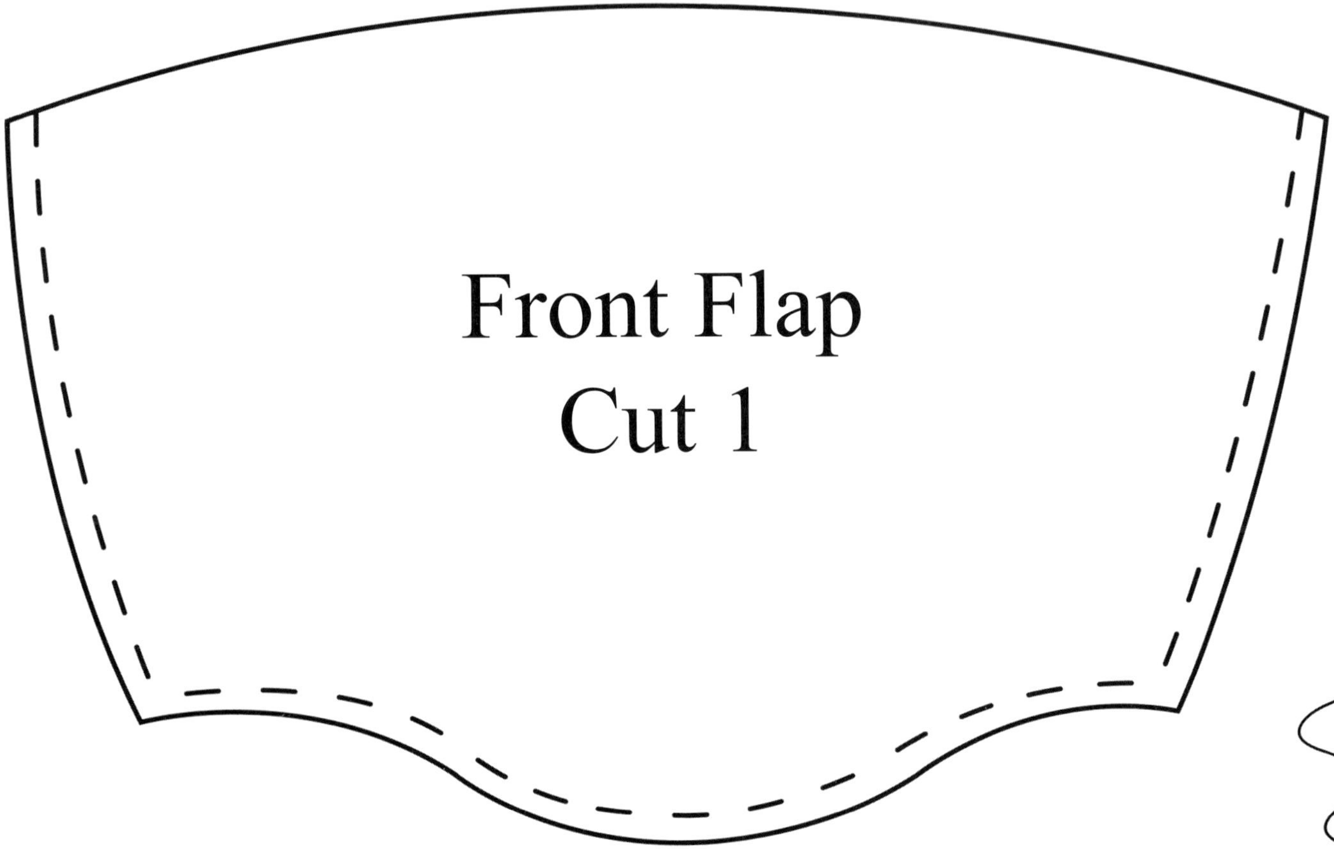

Shown
Actual
Size

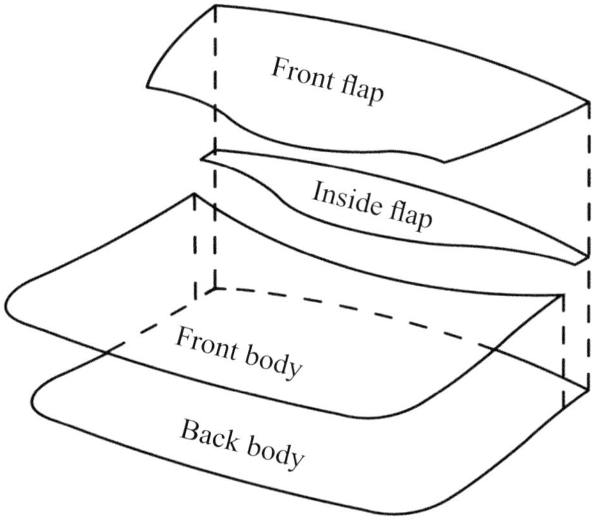

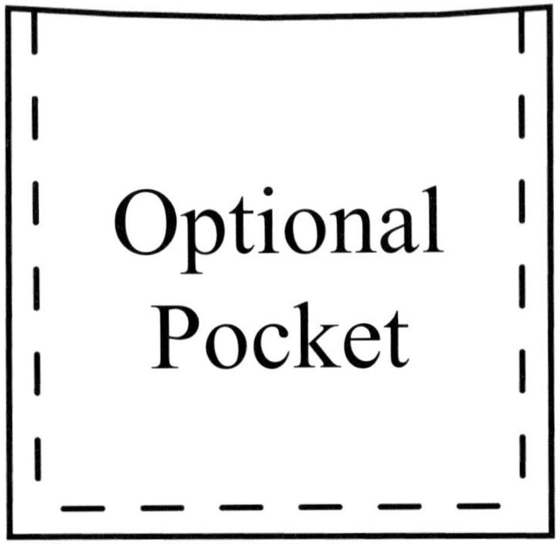

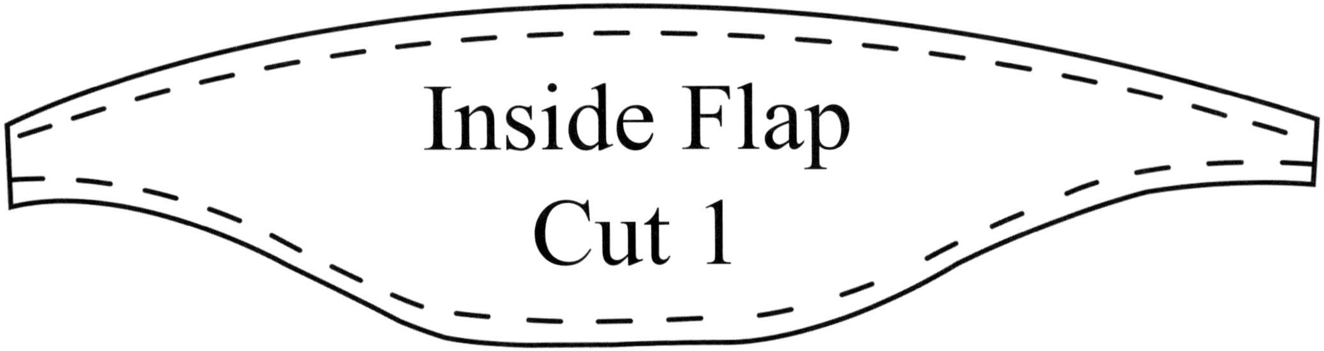

Beaver tail pouch.

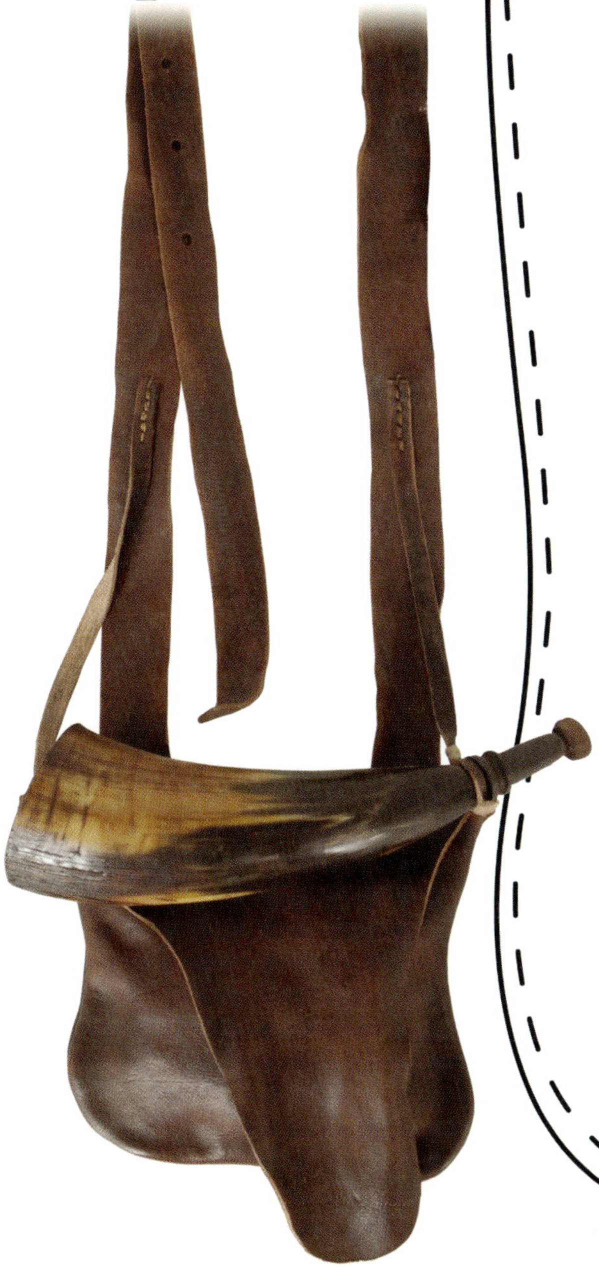

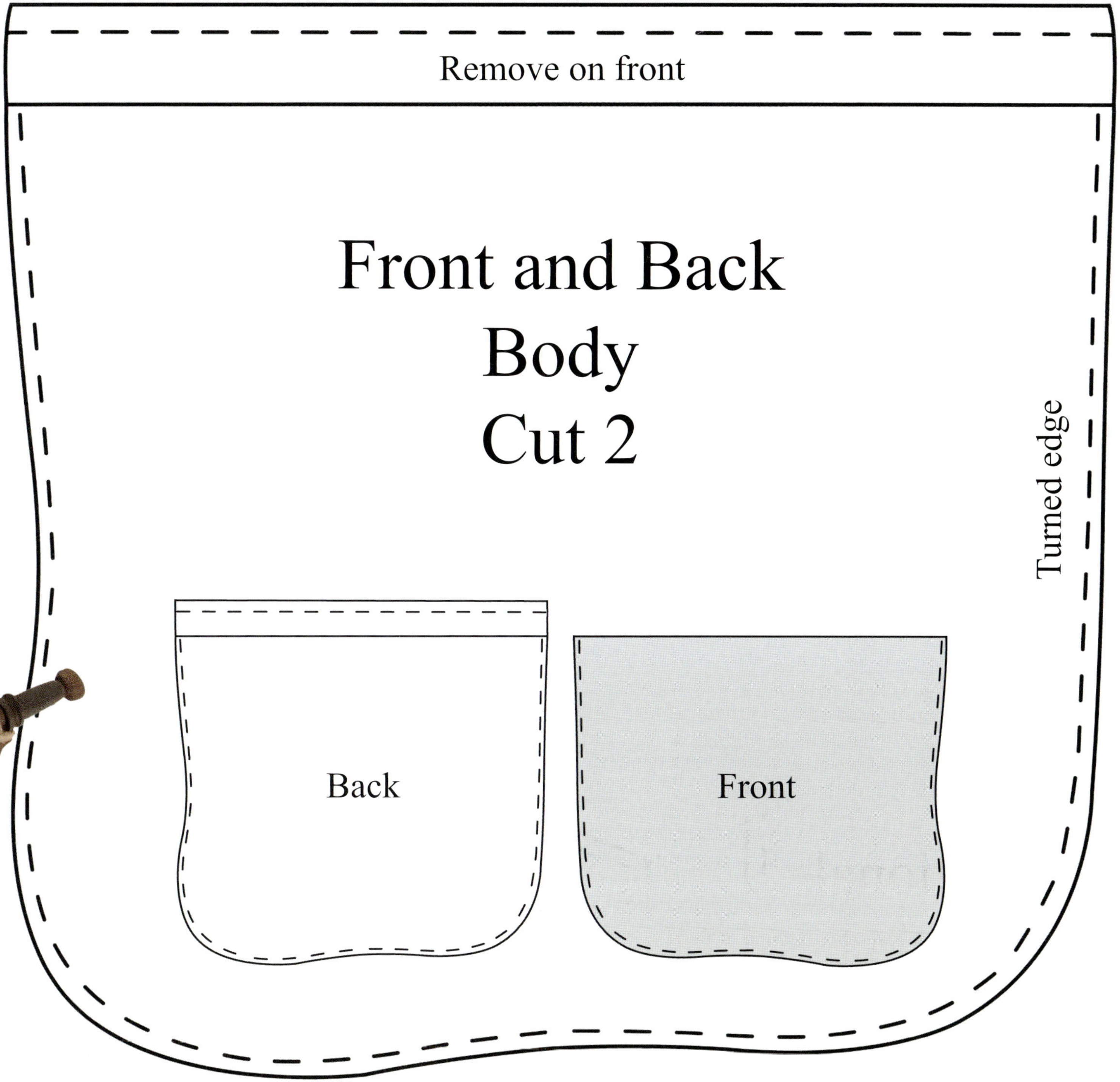

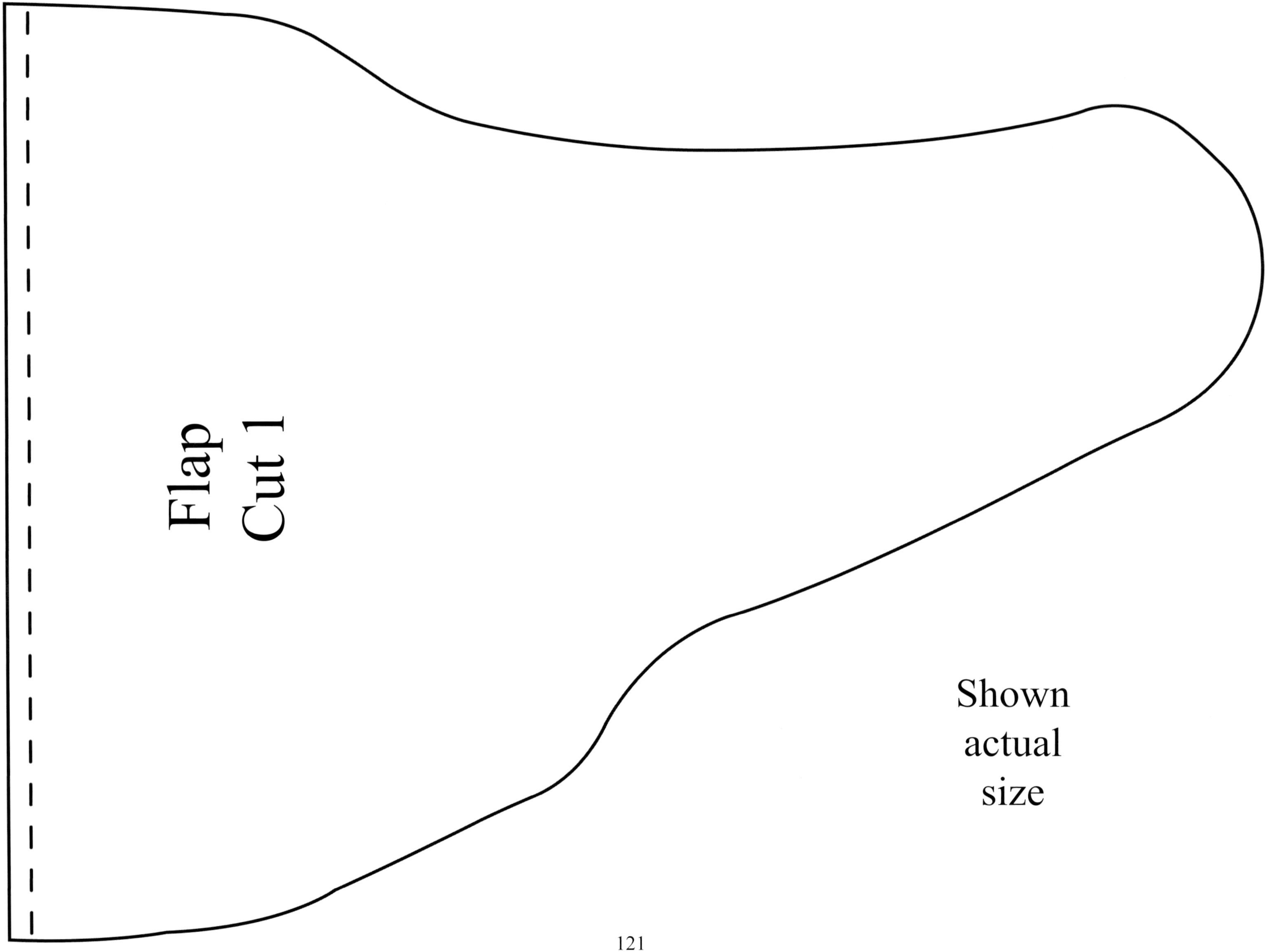

Shown
actual
size

Curved pouch.

original antique pouch
courtesy of A. Rehder

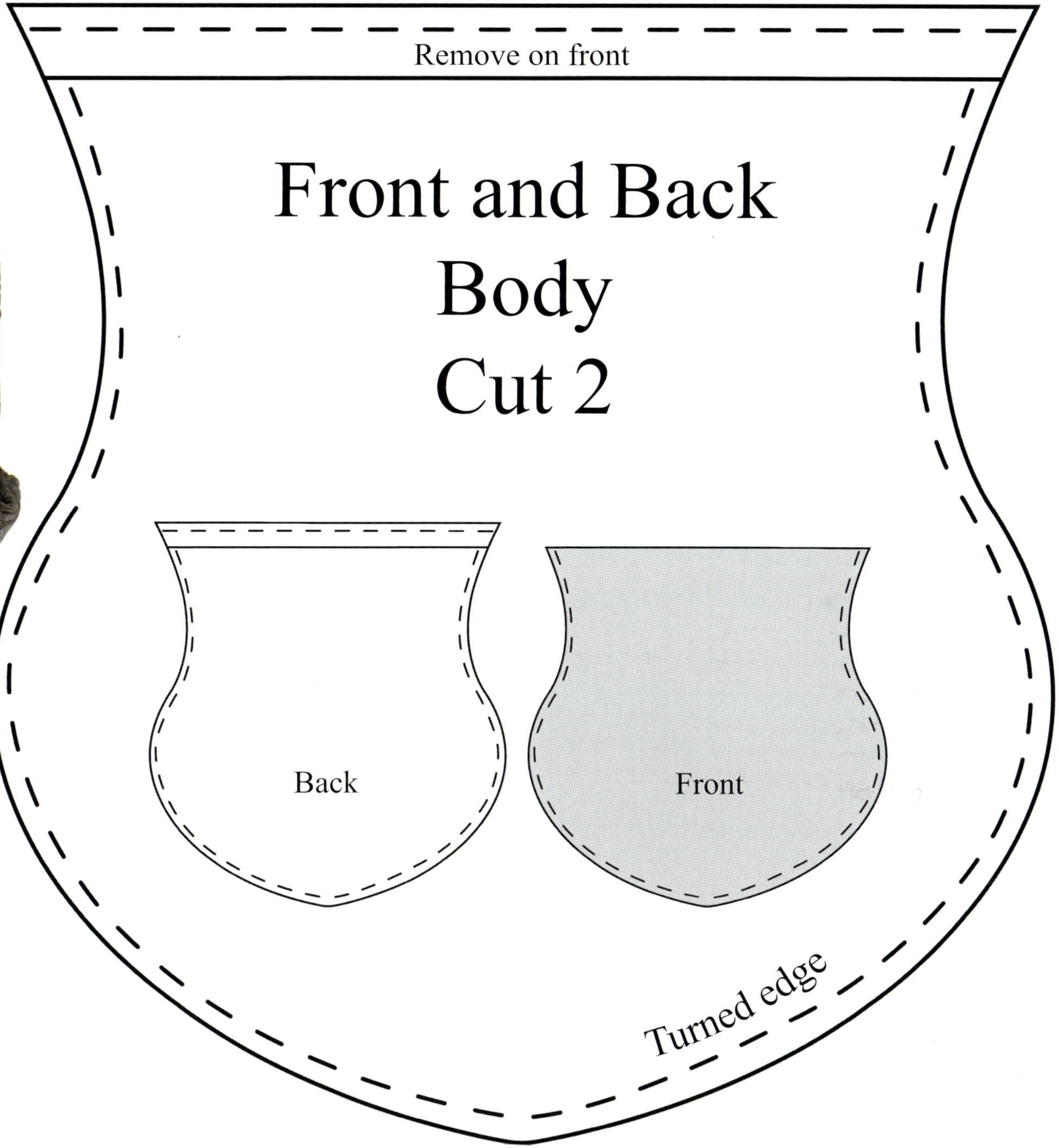

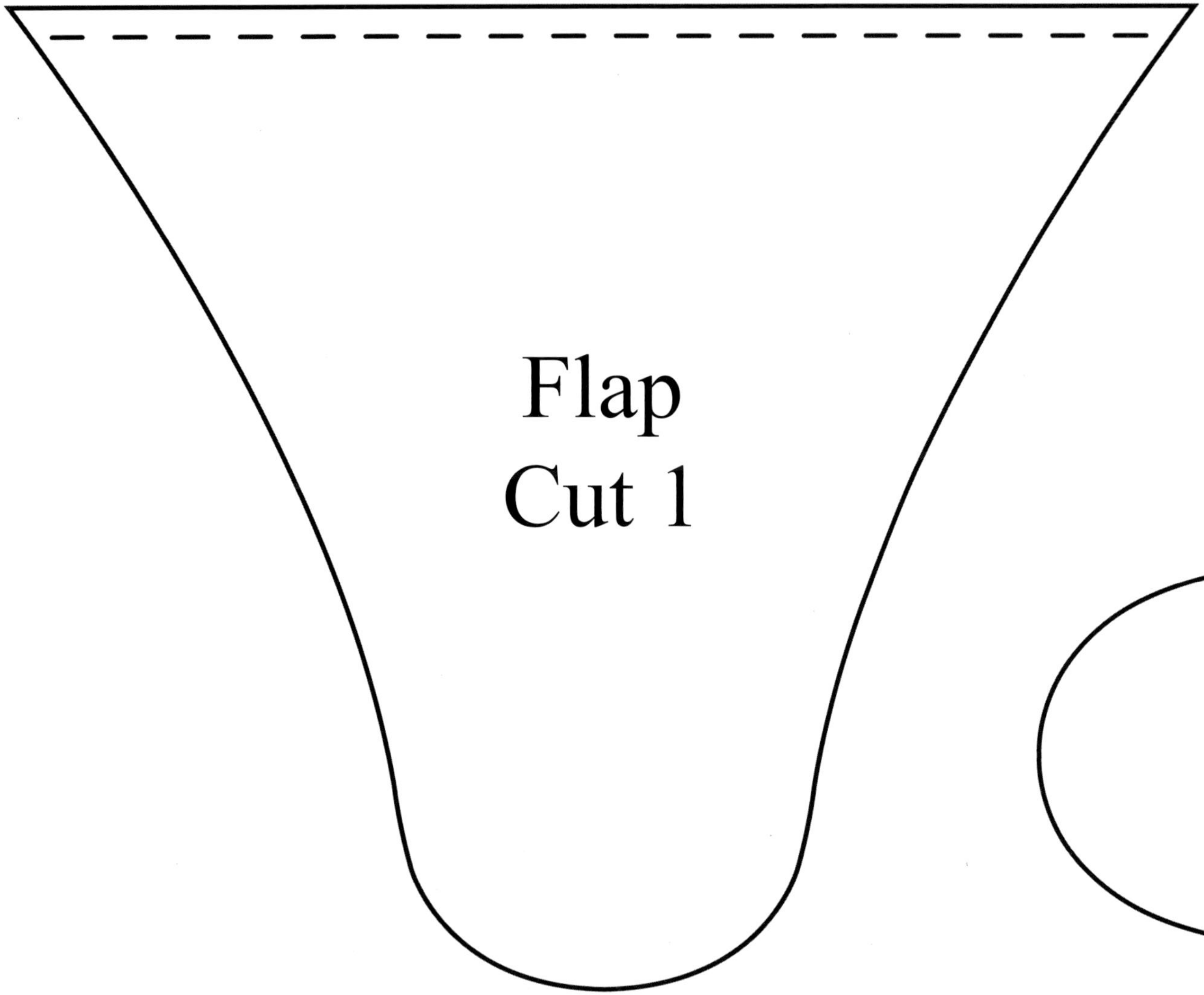

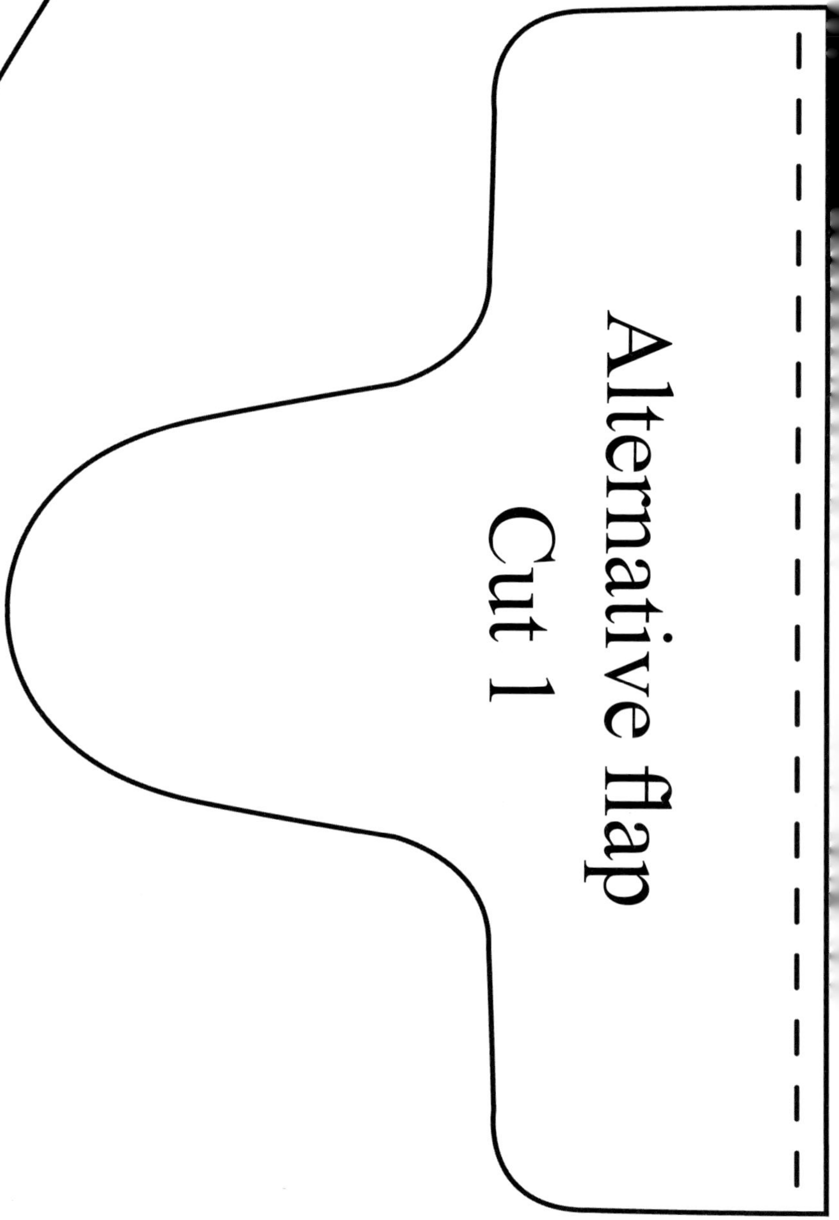

Shown
actual
size